Analytical Chemistry an

Interesting Career

(What Chemists Do)

History, Management and Roles of a Government Analytical Chemistry Laboratory

This book is copyright. Apart from any fair dealing permitted according to the provisions of the Copyright Act, no part may be reproduced, photocopied, stored in any type of retrieval system, or transmitted by any means or process whatsoever without the prior consent of the author or publisher. Copyright is vested in the author. Moral rights are claimed. Every attempt has been made to contact the copyright holder of material used in this book and persons named. Where an attempt has been unsuccessful, the publisher would be pleased to hear from the copyright owner so that any omissions or errors can be rectified. For continuity the uncontactable person's first name and initials of their last name is used. All reasonable attempts at factual accuracy have been made. The publisher accepts no responsibility for any errors contained in this book.

Cover image Chemist Elizabeth Christensen counting Blue Green Algae

First published in hard cover 2022

Revised 2nd Hard cover edition 2023

Revised 3rd Paperback edition 2024

© Gary M Golding

ISBN 978-0-6455451-4-2

First edition printed in Australia by Australian Business Printers Pty Ltd Beenleigh, Queensland

Reviewed "Chemistry In Australia", December 2023 to February 2024 by Bruce Graham FRACI

This is an unusual book it traces the history of the laboratories through the careers of its managers and directors from 1873 to 2023. This format serves also to illustrate one of the themes – that of chemists as problem solvers who adapt to the constant change in the techniques and fields of work.

Anecdotes are also plentiful some chemical some personal...... One described how in the early days one had to give back the empty Biro to get a new one.

Any reader would be impressed by the account of extensive involvement of the laboratory in external activities, including committees, national and international collaborative studies, standard-setting and staff training. This kind of backroom work is often not well recognised.

Commissioner Walter Sofronoff KC

In his comment in his final report "*I have gratefully taken this history from Mr Gary Golding's interesting book, Analytical Chemistry: An Interesting Career, self-published 2022* "

Final Report -Commission of Inquiry into Forensic DNA testing in Queensland page 2 Footnote 1

Radio ABC Brisbane "*full of cracking yarns*"

DEDICATION

To the many scientific staff who spent their entire working lives employing analytical chemistry to enable government decisions based upon accurate, reliable, impartial data, generated without fear or favour, in supporting the goals of justice and public health.

To my wife Frances and my daughters, Amy, Millicent, and Kimberley for their encouragement and patience in supporting this work.

CONTENTS

PREFACE 1

ACKNOWLEDGEMENTS 3

SCIENCE IN THE PUBLIC SERVICE - 1905 5

CHAPTER 1: ANALYTICAL CHEMISTRY 8

CHEMISTS IN THE LABORATORY 11

A CAREER IN ANALYTICAL CHEMISTRY 12

CHEMISTS AS MANAGERS 14

CHAPTER 2: DIRECTORS 1873-1960 21

Karl Theodor Staiger 21

Robert Mar 24

Arthur Walter Clarke 34

Show Dogs Die following Bath 37

John Brownlie Henderson FRACI 38

Frank Connah 44

Leon Meston 44

CHAPTER 3: DIRECTORS AND SENIOR MANAGERS - 1960–2023 54

Ivo Henderson 54

David Mathers 59

Trevor Beckmann.. 62

Des Connell FRACI 65

Oil Spills from Ships 71

Don Lecky 71

Paul Geoghegan.. 73

Michael Moore 75

Ron Biltoft 81

Lenore R 84

Alexander (Sandy) F 85

Margaret W 89

Robyn K 91

Greg Shaw... 92

Craig R 95

Paul C.. 97

John Doherty.. 97

Gary Golding OAM FRACI 99

CHAPTER 4: GENDER BALANCE 108

CHAPTER 5: INPUT INTO LEGISLATION............................ 113

CHAPTER 6: LABORATORY BUILDINGS 115

THE COOPERS PLAINS LABORATORY AND STRUCTURE 118

Philosophy of construction 120

Local Graffiti 124

ORGANISATIONAL STRUCTURE CONSIDERATIONS 131

Variations In Organisational Structure Over Time 132

CHAPTER 7: INTRODUCTION OF USER PAYS 1983 146

CHAPTER 8: QUALITY SYSTEMS - NATA 157

CHAPTER 9: AMALGAMATION WITH HEALTH PHYSICS 166

CHAPTER 10: LABORATORY INFORMATION MANAGEMENT SYSTEM 171

CHAPTER 11: THE LIBRARY 176

CHAPTER 12: SPIN OFF LABORATORIES 182

Bombing of the SS Aramac – 184

CHAPTER 13: CHEMICAL EMERGENCY RESPONSE 185

CHAPTER 14: AUSTRALIAN FUTURE FORENSICS INNOVATION NETWORK-(AFFIN) 189

CHAPTER 15: AUSTRALIA PACIFIC FOOD ANALYSIS NETWORK AND TRAINING 198

CHAPTER 16: AUSTRALIAN AND INTERNATIONAL STANDARDS COMMITTEES 204

CHAPTER 17: PROFESSIONAL BODIES 209

CHAPTER 18: WORKPLACE HEALTH AND SAFETY 1998-2013 213

CHAPTER 19: REVIEW 217

CHAPTER 20: OVER 100 YEARS OF STAFF PHOTOGRAPHS .. 221

CHAPTER 21: EMERGENCY CHEMICAL ANALYSIS OF POLITICAL AND MEDIA CONCERN 224

CHAPTER 22: OVERSEAS USERPAYS ISSUES 233

CHAPTER 23: SECTIONAL ROUTINE ACTIVITIES 2012 240

CHAPTER 24: PERSONAL CAREER STORIES 242

Ron Biltoft 242

Geoff Rynja FRACI 255

Graham Craven 262

Gary Golding OAM FRACI 265

Inge Scott {nee Schott} 270

Daniel John Wruck 273

Stored Chemicals 276

Dr Mannie Rathus 281

Time Management 282

Henry Olszowy 284

Mary Hodge {nee Griffith} FRACI 288

Graeme White 293

Stewart Carswell FRACI 300

John Brownlie Henderson FRACI 304

Henderson's work 306

John {Jack} Adamson 309

William {Bill} Neville Carvosso 311

Howard Couper 312

Keith Deasy 315

Methsiri Edirisinghe 320

Dr Peter Culshaw FCS 329

CHAPTER 25: EARLY NEWSPAPER REPORTS ON THE LABORATORY 340

The Government Analyst 1934 340

CHAPTER 26: ACTIVITIES 2012 - 2022 346

Forensic Chemistry - Work changes 349

Illicit Drugs 349

Clandestine Drug Labs (Clan Labs) 350

Equipment 351

Trace Evidence 351

Laboratory Information Management System (LIMS) 352

Inorganic Chemistry - work changes 352

New work gained 353

Research directions and opportunities: 355

Organic Chemistry: 356

Investigative Chemistry 359

Forensic Toxicology 361

Fume-cupboard Flow Rate 363

Chemistry Annual Christmas Party 367

Environmental Health Officers and the Lab 369

Suffocation at Sea 372

CHAPTER 27: RECOGNITION OF CHEMISTRY STAFF AND FORMER STAFF 374

CHAPTER 28: IMPROVING THE CHEMISTRY LABORATORIES 377

CHAPTER 29: SOME USEFUL MANAGEMENT TOOLS 384

Build Teams 384

Trace Elements in Blood and Plasma 391

Commission of Inquiry into DNA Laboratory 2022 391

CHAPTER 30: OTHER GOVERNMENT LABORATORIES..........396

Pieter Scheelings, PhD FRACI 396

CHAPTER 31: THE FUTURE 404

"BE THERE WHEN NEEDED..406

but make sure they know you are there"...........................406

Finally, We Return to Toilet Paper..407

Appendix 1: ACRONYMS..407

Apppendix 2: Authors background 409

Appendix 3: Publications and Conference Presentations....410

PREFACE

"The health of the people is really the foundation upon which all their happiness and all the powers of the State depend."

Benjamin Disraeli 1804 – 1881

This book celebrates the 150th anniversary of the appointment of the first Government analytical chemist in Queensland in January 1873. It is a review of some of the Government analytical laboratories' activities since that time.

The following is the author's personal view of the activities of the laboratories during the period 1960-2013. The views expressed are those of the author and thus cannot consider any activities of which he was unaware. This was a period of dramatic change and growth, where the survival of the laboratory was often at risk.

The author, Gary Golding, commenced working at the laboratory on a permanent basis on 14 December 1971 and continued to work at the laboratory until September 2013. At the end of his career, he was the

Managing Scientist of all the chemistry sections in a vastly expanded multidisciplinary laboratory.

There are very few environmental chemical contamination concerns, criminal activity, public health issues, chemical exposure events, product safety, international trade contamination issues or personal tragedies involving suspicious deaths in Queensland that don't arrive at the analyst's bench. Almost every day the newspapers report our results without acknowledgement. Every day, the courts accept our certificates, allowing them to carry out the course of justice.

Apart from the routine work, a constant stream of unusual samples is received. Some of these cases are described throughout this book, illustrating the laboratory's role as the **laboratory of last resort** for clients with novel problems other labs are unwilling or unable to solve.

Many staff devoted their entire working life to the development of the Queensland Government Analytical Chemistry laboratories. Unfortunately, there are few records of their achievements. This book hopes to list just some. The author also hopes that future managers will use this book as a reference to what has worked and what has not worked in the past and perhaps gain some insight into the reasons for the laboratory's existence and its value to the State and community.

Careers in analytical chemistry can offer a work life full of variety if one chooses. The author has often commented that he never seemed to carry out the planned activities of the day due to some unforeseen government or court priority.

The laboratory has had several names over the past 150 years. Recently Analytical Chemistry has been one of the disciplines within these entities. It is not the intention to discuss the history of the larger organisation. For simplicity, the term Government Analytical Chemistry Laboratory has been used as a generic name to cover the various name changes the organisation has had.

ACKNOWLEDGEMENTS

The author wishes to thank Don Lecky, Ron Biltoft, Paul Geoghegan, Geoff Rynja, Pieter Scheelings, Mary Hodge, Graeme White, George Lee, Peter Culshaw, Des Connell, Graham Craven, Inge Scott, Stewart Carswell, Eva Comino, Lenore Hadley, Henry Olszowy, Methsiri Edirisinghe, Tatiana Komarova, Robyn Mackenzie and Dennis Burns (Australian Future Forensics Innovation Network) for their contributions and support. The author especially thanks Trish Murphy and Alyson McCulloch for their work in editing, design, and publication of this book. Special thanks to Peter Culshaw for his research on newspaper items related to the history of the laboratory. The author also thanks Stig Hokanson for his advice on publication. Photographs of past directors and staff, courtesy Queensland Health Forensic and Scientific Services.

A special thanks to Primo Levi for his book, "The Periodic Table," about the life of an analytical chemist, which inspired this project.

The author thanks Queensland University Press for their approval to quote from "A History of Health and Medicine in Queensland 1824 to 1960" by Ross Patrick (1987).

SCIENCE IN THE PUBLIC SERVICE - 1905

The Brisbane Courier, (18 November 1905 page 12)

"There is no department in the public service so rich in material for the sensational writer as the Government Analysts department. The groundwork of many exciting plots might be taken from some of the investigations; and, by peculiar coincidence, much that is comical, and much that reveals the gullibility and fads of the public passes in review. The analyst deals in tragedy, comedy and sometimes farce; only his materials speak through their constituent elements rather than from behind a mask. In cases of murder, it is his scientific skills which are often employed to identify suspected evidence of guilt, and many a perplexing crime has been made clear by chemical analysis. The lighter shades, so to speak, are supplied by analyses of famous nostrums (remedies) which the public believe have a sovereign virtue exceeding by far the magic of the touch of a kingly hand, but which, giving up their secrets to the analyst, stand revealed as miserable shams. Even comedy may broaden into farce, but some of the strange ideas of the unenlightened are enquired into. The government analyst is really the secret police of the country, and he stands sentinel for the public, guarding them, whenever he is permitted against insidious and well-cloaked crime, against fraud in high and low places and against adulteration, and not infrequently protecting the public against themselves. A wonderfully interesting and very important department is this, if one had the pleasure and the opportunity to fully study its works and its records. Author comment; **Read on**.......)

In the twenty-first century the Queensland Government's Analytical Chemistry Laboratory's mission remains unchanged; to protect the public from those who through greed, negligence, or criminal intent, would damage the health and well-being of others."

Chemist Geoff Eaglesham summed up the life of an analytical chemist with this organisation in his retirement resignation letter.

"The time has come for me to bid farewell to QHFSS. I have enjoyed my working life with this organisation, and it has presented plenty of variety to keep me interested in my job for the last 44 years. That's a long time with one employer but with ever improving technology and a new analytical challenge seemingly every few months, I can't say the job has ever been boring."

CHAPTER 1: ANALYTICAL CHEMISTRY

"A man who qualifies himself well for his calling, never fails of employment."

Thomas Jefferson 1743 - 1826

Modern society runs on measurements. Analytical chemistry is the application of tthe chemical sciences to identify and quantify substances to generate data/measurements for decision making purposes. It is the science of chemical measurements but also includes elements of physical testing of the properties of items e.g., viscosity, strength. It involves the use of various techniques and instruments to identify and determine the composition, structure, and properties of chemical substances and their mixtures. Analytical chemists develop and apply methods to separate, identify, and quantify the components of a sample. This field plays a crucial role in various scientific and industrial applications, such as environmental monitoring, drug testing, forensic analysis, quality control, and research and development in diverse fields. It is one of the silver threads that binds societies together through the generation of measurements that all parties can accept.

The aim of the analysis is to test the item for conformity to a specification or regulation. It is incumbent upon the analyst to ensure their results are as accurate as possible. Multimillion dollar contracts rely upon the results of the chemist's test e.g., sulphur in export coal. National income can be diminished should an exported product not meet a specification, e.g., pesticide levels in exported meat. The health of individuals is dependent upon the integrity of medical measurements. The freedom of individuals can be affected by the results of forensic analyses, e.g., the level of sentence often relies upon the amount of drug present. Control of manufacturing processes relies heavily upon measurements made on the factory floor e.g., for consistency of product. The results of research and its reproducibility by others require accurate analytical measurements.

Getting the measurement result right is not an easy task. The project management parameters of time, quality, and cost come into play. Quick, cheap, and right can be a challenge.

A method that works for one product may not be able to be applied to another because of different chemical characteristics that cause an elevation or diminution of the data signal being measured. Methods that detect and quantify very small quantities may not provide accurate results for percentage quantities.

The work of an analytical chemist can cover a huge scope of issues from the personal to the international. Several examples have been outlined. All are actual cases the author and colleagues have encountered during their careers.

GOVERNMENT TOILET PAPER

Up until the early 1990s, government toilet paper in Queensland was of a unique design. One side was shiny with a mirror finish and the other side roughly finished with the consistency of fine emery paper. If held up to the light, the brown-coloured paper was transparent. The paper had almost zero absorbency for moisture. The contract for this paper dates to antiquity as presumably did the formulation. It is unclear why the government chose this type of paper, but I suspect it was to prevent staff stealing it. I am confident the strategy was very successful. Staff bought their own rather than use it. In the UK, a similar paper could be found in government premises with each sheet marked 'government property.' In the 1980s the laboratory, as part of a tender process for supply of toilet paper, rejected this paper as unsuitable for human use. The result was that the contract was let to a company that made paper to a more modern design. All government departments and their staff can be thankful for the efforts of the Queensland Government Analytical Chemical Laboratory in removing this item from State facilities.

CHEMISTS IN THE LABORATORY

"The greatest aim of education is not knowledge but action."

Herbert Spencer 1820 – 1903

Once you have graduated, a career in analytical chemistry usually starts out as a process chemist performing routine operations in a laboratory.

This work can be very routine but will expand the chemist's knowledge. Over time with personal development, the chemist will move on to more challenging work. This could include non-routine samples, troubleshooting or a role as an instrument "jockey" specialising in performing analyses for others using a state-of-the-art instrument. The more experienced chemist may be entrusted with work of a legal nature.

There are four basic laboratories the chemist may work in. A government analytical laboratory which carries out a wide range of work. A commercial analytical laboratory which deals mainly with high-volume short turnaround time samples. The industrial laboratory monitoring product production parameters. A lot of the production testing has been removed from the centralised quality-control laboratory to the factory floor next to production. This significantly reduces the time for the analytical work to be performed. The industry laboratory may not employ a lot of chemists. Often, this work is performed by unqualified factory workers who have been trained to carry out a particular test. Others include a university laboratory which carries out research. The research laboratory will have a lot of equipment that requires specialised operators. A lot of this equipment will not be fully utilised but will spend most of its time idle. However, it provides a capability when the need arises. In contrast the commercial laboratory must justify the purchase of equipment based on sample through put rather than the need as part of a research project.

A CAREER IN ANALYTICAL CHEMISTRY

"Genius is mainly an affair of energy."

Matthew Arnold 1822 – 1888

The first step in becoming an analytical chemist is to have a general interest in chemistry. The next step is to gain the correct qualifications. There are many courses which purport to lead to a career in analytical chemistry. Many of these focus on forensic science or environmental science. However, these courses contain a lot of subjects which are not relevant to the analytical chemist. Any reduction in chemistry content in the course will reduce the level of your knowledge and thus competence. It is important that the third year of the course has a heavy emphasis on chemistry. It is also important to note that doctorate courses, following a non-chemistry degree, which involve some chemical analysis may not lead to employment as an analytical chemist. Once again if you want to be an analytical chemist you should study analytical chemistry and not the plethora of other subjects which although useful and do involve some chemical analysis, do not give you the full education in this area.

The public will confuse your job with that of a pharmacist, "Where is your shop?"

they will ask. Tell them "I am a **real** chemist not *a pharmacist*".

SAMPLING IS IMPORTANT

Failure at this stage renders the results invalid. For example,

determination of the moisture content of petrol. This was usually done by the Carl Fischer technique. A sample was pipetted from a drum. The result seemed quite high (0.3%). The sample had been taken directly from the metal container. When the petrol was emptied out it was discovered that the contents were at least 50 percent water as a separate layer under the petrol. A lesson in ensuring the sample is homogeneous.

Forensic analysis is one of the few areas where you cannot specify the sample size. You may not be allowed to use all of the sample. Some may need to be retained as evidence. Often you must do what you can with what is provided by the police. Here judgment comes into play. You will be amazed how much you can do with a very small sample.

CHEMISTS AS MANAGERS
"When we think we lead we most are led".
Lord Bacon 1788 – 1824

Within the laboratory and in industry you will be required to manage other staff. This may initially be just one technician, a factory worker or a less experienced chemist. This process builds managerial skills, and it is not uncommon for industrial chemists to move into a production manager's role. Once the transition into management occurs, opportunities for promotion into other areas of the organisation open up. Chemists are trained problem solvers. During their scientific training they must answer thousands of problems presented to them by the examiner. This training will stand you in good stead in a management role.

An alternative to a management role is to become a specialist researcher and devote your time to building expertise in a particular area of chemistry. This may be difficult to do in a factory laboratory. Universities provide a venue to achieve this skill base. Over time, regardless of your direction, you will become a trainer, mentor, and adviser to others.

If you work within government, try to get on some committees. You may not have much knowledge of the area but some quick research or reading a book on the subject will soon make you more knowledgeable than most of the other members. Once again, your training in problem solving will make you a valuable addition to the committee. The author was once on a committee looking at the proceeds of crime. In this case a house that had been used as a clandestine methyl amphetamine laboratory. The meeting was bogged down trying to avoid demolishing the entire building. The author asked where in the house the lab was located. It was in the laundry. It was known that this house was old and in the past the laundry was often built onto the back of the house so could be demolished without damaging the rest of the house.

One role is to become a consultant. You will be asked to investigate particular problems. This may involve taking samples. This role is less capital equipment intensive than a full laboratory role. Samples you collect will be sent to a particular laboratory you trust for analysis. With the advent of instrumentation which can carry out tests in the field it is often advantageous for the consultant to purchase or rent a particular piece of equipment for a job. This role could possibly take you to other parts of the world. It is possible to specialise in certain areas. You may wish to provide testing and advice for insurance companies as part of arson investigations or paint defects. The role of the chemist crosses over into areas of environmental consultancy. This can include contaminated site clean-up or the assessment of releases of contaminated water into the environment from mine sites.

Continuing education is the key to success. It is advisable at some stage in your career to diversify your qualifications by adding initially a higher degree in chemistry and later a business or management degree to your resume.

You should join a professional body such as the Royal Australian Chemical Institute (RACI). This will enable you to develop your skills managing projects, including local seminars and international conferences. It will also build your network of collaborators and possible employers. Your public speaking skills will be developed through managing RACI meetings and through conference presentations.

Research has shown that people who say "yes" to the manager when you are asked to do something different, tend to be more successful in the long run.

This book looks at the Analytical work performed by the Queensland Government laboratories, and the work is remarkably consistent over time. The reason for the consistency is that it supports the main role of government, a secure society, a healthy society, an educated society, and an economically prosperous society.

Later in your career as a manager you may return to the once-familiar lab and find it feels foreign. Such is the nature of analytical chemistry that it is always changing. However, the reasons for the work, in the case of a government laboratory, remain the same although the techniques used vary over time.

By reading this book you will get an insight into the work of several analytical chemists, the different stages of their careers, and gain some knowledge on the common analyses performed. It will also give you an indication of the scope of the work, which is very large. The examples given are taken from the careers of chemists known to the author. In any workplace there is a degree of organisational politics, personality clashes, and competing ambitions. You need to be aware of this aspect of working life and adapt accordingly. Some managers are consultative, and some are not consultative. These aspects are not covered in depth in this book. Many of the laboratory's achievements go unrecognised, some major, some routine and some with a personal impact.

BREAD WITH UNUSUAL FILLING

A loaf of bread was received with a rat baked into the loaf.

SAVING THE AUSTRALIAN WHEAT CROP

Gary Golding received a phone call from a person requesting analysis of a section of blue tarpaulin sheeting. Apparently, the product had failed in some areas and was allowing water through. His first thought was that it would be less expensive to buy a new tarpaulin. When asked what the sheeting was used for, the caller indicated it was used to cover the Australian Wheat crop. The value of the Australian Wheat crop in 2012 was $US7.6 billion. Microscopic examination showed that the pigments were not uniformly mixed. This resulted in a deficiency in UV inhibiting chemicals in some areas. When exposed to sunlight these areas failed. Microscopy is a powerful analytical tool.

DUSTY CARPET

A lady rang the laboratory wanting us to test the contents of her vacuum cleaner. Apparently, there was so much dust in her new carpet that it rapidly filled the vacuum cleaner bag. She had been to several laboratories and made complaints to the carpet retailer. The manufacturer could not help. An infrared spectrum of the dust matched Scotch Gard a product used on carpets to repel dirt. It was obvious from these results that too much Scotch Gard had been sprayed on the carpet during the manufacturing process. Armed with this knowledge she was able to get the manufacturer to replace the carpet.

BABY WITH RED BOTTOM

A nappy, provided by a nappy service, was submitted along with a complaint that it had burned the baby's bottom. A quick pH test indicated high acidity. Chemist Christina Malar evaporated a water extract of the nappy which yielded crystals. Infrared analysis of the crystals indicated oxalic acid. We knew from our Government Laundry Committee work that oxalic acid was used in commercial laundries. Investigation of the laundry at a prison, which was contracted to wash the nappies, found that the inmates were soaking the nappies in oxalic acid then throwing them into the dryer without washing them

SCHOOL CHEMISTRY MAGIC

Through the RACI the laboratory staff visited local primary schools to provide a magic chemistry demonstration. These included exploding hydrogen balloons, liquid nitrogen displays, chaotic equilibrium with starch iodine, paper chromatography of black felt pens, changing copper into silver, and microscopy with crossed polarised light viewing crystal formation as the sample solidified on a hot stage. Following one of these demonstrations one of the children passed the author in the street and said he was going to be a scientist when he grows up.

ASBESTOS IN BUILDINGS

Until the late 1980s it was common for buildings to be constructed using an asbestos cement sheeting (Fibro). The laboratory routinely received samples of fibro to determine if it contained asbestos. The analysis was performed by microscopy.

CHAPTER 2: DIRECTORS 1873-1960

"Quick decisions are unsafe decisions."

Sophocles 49 5-406 BC

The early history of the Queensland Government Analytical Chemical Laboratory, until 1960, is outlined in the book "A History of Health and Medicine in Queensland, 1824 to 1960" by Ross Patrick (1987) Queensland University Press. A summary of this period taken directly from this book follows, with editorial review.

In 1874 James Dixon, the member for Enoggera, asked the Colonial Secretary Arthur McAllister whether the government intended to make any provisions on the 1875 estimates for a Government Analytical Chemist. The Colonial Secretary replied that provision had already been made in the estimates for the salary of an analytical chemist at present employed by the Government (Patrick, p. 134).

KARL THEODOR STAIGER

Government Analyst 1873-1882

Photo Courtesy Qld State Library

Karl Staiger had been appointed in January 1873 to act in conjunction with the police force in detecting parties engaged in the manufacture and sale of imported adulterated food and drink. The 1872 Queensland Health Act, dealing solely with infectious diseases, was silent on the matter of food and drugs (Patrick, p. 134). Three New South Wales Acts adopted by Queensland dealt with the prevention of adulteration of food products and liquor. To support these Acts, in January 1873 Karl Theodor Staiger was appointed. He was attached to the Department of Public Works, later the Department of Public Works and Mines, and when the estimates were being prepared his salary was recorded alongside the expenses related to gold mining (Patrick, p. 134).

The laboratory continued to provide a service to the mining industry in gold and silver assays well into the 1980s. Samples were analysed for the Cloncurry Assay Office for copper, gold, and silver every week, as a free service to holders of Queensland prospecting licences.

The media has a long history of reporting on the activities of the laboratory. In September 1881, the Brisbane Courier commented upon a report entitled 'Analyses of Spirits, Groceries et cetera'. The report contained the results of examinations conducted by Staiger of samples from houses and shops in Brisbane and Queensland country towns. The Brisbane Courier expressed the opinion that the results of the analysis indicated a need for the immediate passage of the Adulteration Bill before the house.

The Brisbane Courier.

PUBLISHED DAILY.

Tuesday, September 13, 1881.

A PAPER entitled "Analyses of Samples of Spirits, Groceries, &c.," laid before both Houses of Parliament last week, affords timely information on the subject of adulteration. The Assembly will probably consider the Sale of Food and Drugs Bill in Committee of the Whole during the present week; it will be well, therefore, for honourable members as well as the public generally to direct attention to this important subject.

The paper contains two analysts' reports, one signed by Mr. J. Cosmo Newbury, of Melbourne, perhaps the best Australian authority on the subject; the other is signed by Mr. C. T. Staiger, the late Government analyst in this colony. The samples were, it is understood, procured from various public-houses and shops both in town and country for the purpose of analysis, with a view to ascertain whether or not the popular im-

The Bill was given Royal assent in October 1881. Under its provisions, the Governor-in-Council was empowered to appoint a person processing competent knowledge, skills, and experience as government analyst for the purpose of the Act (Patrick, p. 134- 5). More recently the appointment of State Analysts was delegated to the senior director of the laboratory. The duties of those appointed were to be defined by regulation' (Patrick, p. 135). This approach continues to this day with the appointment of State Analysts under various Acts. Tracking the people appointed under the various Acts was necessary as each Act came into force. The process was simplified in earlier times, when some Acts simply stated that the analysis was to be performed by an analyst appointed under the *Health Act*.

'The general execution of the Act was placed in the hands of the local authorities who were empowered and if required by the Minister, compelled, to appoint a Public Analyst' (Patrick, p. 135). It is interesting to note that although Mar was the first official Government Analyst, Karl Theodor Staiger was referred to as the Government Analyst in this newspaper report from 1881. In 1888 Staiger died of tuberculosis in his home, which still stands.

ROBERT MAR

Government Analyst 1882-1892

"Persons with the necessary qualifications were hard to find in the colony and the government arranged for Robert Mar to come to Queensland" (Patrick, p.135).

The Telegraph 10 May 1882 stated that: "Amongst the passengers aboard the Chyebassa is Mr Robert Mar, the recently appointed Government Analyst. Mr Mar studied chemistry at the University of Glasgow under the late Dr Thomas Anderson, and afterwards under Professor Ferguson, who recommended him to his present appointment. Mr. Mar, who is quite a young man, comes here with high credentials as to his scientific attainments. " Customs, Excise and Police Departments made the most demand for his services. For the Customs Department, he analysed samples of tea and determined the flashpoint of kerosene. In his work for the Excise Department, his analyses were used to assess the adulteration of liquor (Patrick, p. 135). With the reduction of tariff barriers in the 1980s customs work gradually diminished.

Proceedings of the Royal Society of Queensland 1955 (volume 67) reported on Mar's work: The good citizens of Brisbane were not entirely without creature comforts as can be seen from the fact that aerated waters, cordials, and rum were manufactured here as early as 1852. As a matter of interest, a Brisbane firm, Dark and Stalker, won the first prize at the Sydney Show in 1879 for its ginger beer. However, we should also note that a careful check was soon to be kept on such things by the Government Chemical Laboratories. We shall hear more of them later, but from the first report of the first Government Analyst, Robert Mar, which was written in 1883, we learn that:

'From the samples of such liquors thus far submitted to me, I judge that any pernicious effects, consequent upon the use of those sold in Brisbane, are due to the spirits themselves being too new, and unmatured, and not because of adulteration with foreign, injurious substances. The vendors reduce with water, colour with burnt sugar, and, in some cases, add a little flavouring matter; but, in the samples examined, water only has been found in excess. Four adulterated milks came from Townsville and three (of the six) adulterated whiskies from Warwick. *The Queenslander* 17 Oct 1885, page 634 gave a synopsis of Mar's activities in an article titled, Scientific and Useful:

From the report of the Government Analyst, Mr Robert Mar FCS, for the year ending 31 July last, it appears that there were 360 samples dealt with on account of various Government departments. There were also 202 analyses and determinations in connection with the foregoing making a total of 562. We take the following from the report: of the 107 teas examined, thirty were adulterated, and six of these reported "unfit for human food". The adulterants found were exhausted tea leaves (from 1 to 10 percent), foreign leaves, starch, gum, catechu [extract of acacia trees], magnetic oxide of iron, clay, and sand. The sixty-six samples of kerosene represent eight shipments of oil, and of these, six I found flashed below 110°F.

The wine was unadulterated, and no adulterant was detected in the beers. Of the spirits, four had been reduced with impure water, and an excess of water (from 2 to quantity of petroleum; three rums contained from 116 to 126 percent of proof spirit and were too new for consumption.

One of the six teas condemned as "unfit for human food" contained only about 50 per cent of genuine tea. The other five were less sophisticated, and in these cases I (Mar) added to the results of examination and analysis, the considerations from physiological experiment—upon myself before making a report. But this is not a convenient method for confirming an opinion, and yet it is the only one possible in doubtful cases. *Testing products on analysts still occurred in the 2000's where a physiological reaction had been reported by the client.* I (Mar) therefore suggested that those teas should be prohibited from going into consumption which in quality fell below the British Public Analysts Society's limit, as is the case in New Zealand. This suggestion has been acted upon, and an extra officer of the Customs (whom I instructed, and who has shown aptitude in apprehending and skill in following the instructions given has been deputed to make a preliminary examination of all teas coming into Brisbane. The result of these arrangements I have tested by making analyses of teas imported in previous years and comparing the same with those of teas presently coming into Queensland and finding the proportion of adulterated tea sent has materially decreased, and a sensible improvement has taken place in the quality of non-adulterated teas.

Several large shipments of kerosene having within the last few months been condemned at the Customs here, dissatisfaction has been expressed with the test prescribed by the *Mineral Oils Act* of 1879. It does not state how long the latter should take and does not say whether the lowest temperature at which a flash was got, or the mean temperature of several tests should be regarded as the flashing-point of an oil. I have made comparative experiments, the results of which I have already detailed in a separate report — with the open and closed cup tests and I am entirely in favour of the adoption of the latter as the method to be used for testing oils imported into this colony. If, as has been supposed, the adulteration of spirituous liquors with noxious drugs has in past years been extensively practised in Queensland, it is, I think, evident that such practise is rapidly decreasing, at all events in those districts from which samples have been taken. And here it may be said that in Britain the adulteration of spirits with substances other than water is of rare occurrence, and as American analyst (Dr. Englehardt) states, regarding whisky, that it appears "evident that the addition of water and colouring matter is more practised than any other adulteration" and, regarding rum, that "no injurious foreign substances were detected. The sample of gin in which strychnine was found came from a private individual through Mr Waters, inspector of distilleries; but two other samples of gin from the same public house, and said to be the same spirit, which were procured by another person, contained no trace of poison. Many toxicological examinations for the Police Department were conducted by Mar to determine the cause of death, as poisons were often used. The stomachs sent for analysis were mostly those of persons who had committed suicide, and it appears the facilities are too numerous for obtaining quantities of deadly drugs. It is doubtless due to the more common use of strychnine for getting rid of troublesome pests that such poison is most in favour of self-destruction. However, it is remarkable that persons who, in all probability, are well acquainted with the effects of the same do so choose, than poisoning by it, a method more painless of killing themselves.

In 2012 this work still formed the mainstay of a large volume of work in toxicology. With increases in sensitivity of instruments, blood is now the preferred sample for drug screening. The work has been extended to include saliva testing for drugs in drivers and later drug testing in hair.

Mar had the satisfaction of seeing the *Sale of Poisons Act* passed in 1891. After 10 years, the position was terminated at the end of 1892. While he was well occupied, there is little evidence to show that his services were used to any great extent to implement the legislation which had brought about his appointment (Patrick, p. 135).

Laboratory staff have provided technical advice to health inspectors and environmental health officers on drug and poison committees continuously since this time. When the Federal Government introduced uniform scheduling of drugs and poisons, the Commonwealth laboratories took over a regulatory testing role. The work in this area dropped considerably, and only samples of an unusual or non-routine nature were submitted by Environmental Health. For example, samples checking for the pilfering of Schedule 8 medicines by hospital staff was not an uncommon request.

TEXTILE DYES IN IMPORTED LOLLIES

In the 1990s, when the colourings in some imported lollies were examined. It was found that textile dyes were used. The laboratory routinely screens food and drinks for artificial colouring. In the past paper chromatography was used to confirm the identity of the dye. Now HPLC techniques are available. F.E. Connah in 1937 said the US has 15 permitted food dyes, Qld had 27.

NEEDLE STICK INJURIES

One of the major hazards of testing syringes suspected of containing an illicit substance was the chance of needlestick injury. In the early 1980's, the samples were delivered with inadequate protection for the analyst. Often needles protruded from the sides of brown paper envelopes. With the increased concerns about both AIDS and hepatitis B infections, consideration was given to suspending this type of work due to the impossibility of safely performing the task. This corresponds to the first level of hazard control, i.e., stop doing the hazardous activity. However, analysis continued. Various techniques were applied including cutting the needles off syringes. This once again involved handling the syringes. On occasion the cut needle would fly through the air. Packaging of these materials has improved significantly.

Champion Draughts Player Dead.

The sudden death has occurred at the Diamantina Hospital of Mr. Robert Mar, the champion draughts player of Australasia. The deceased had been in ill-health for some time, but the immediate cause of death was hemorrhage. The late Mr. Mar was at one time Government Analyst in Queensland. He wrested the title of champion of Australasia from Mr. Warnock about seventeen years ago, and retained it up to the time of his death.

Death of Robert Mar Brisbane Courier 22 Dec 1908

ALWAYS QUESTION THE RESULTS

A technician verbally reported that there was no chromium detected in the blood sample. The follow-up question was critical. How did the standards go? The answer was... "there was no chromium in them either!" On another occasion a methyl barbiturate was detected by GCMS. The case involved

a suspected attempted murder. This result was very significant. The author was aware from his experience working in toxicology that this substance was not commercially *available and was used as an internal standard for quantification of barbiturates via on-column methylation. Investigation showed that the previous sample on the instrument was a methylated barbiturate and that this result was carry over. A great injustice was avoided by a broad generalist knowledge.*

With experience you will develop a bell in your head that rings when you see a result that is wrong. Do not ignore the bell.

GLASSWARE WASHING

In the days of wet chemistry, a large amount of glassware needed to be washed. Most sections had a dedicated glass washer who took care of this. Upon the move of the laboratory to Coopers Plains the glass washing facility was centralised. Dishwashers were purchased. This reduced the dermatitis caused by continual contact with water while hand washing glassware and cuts from broken or chipped glassware. In one particular area, on a Friday afternoon, the sound of breaking glass became more apparent possibly related to the lunchtime Friday pub visit. With centralisation of glass washing, at one point, the glassware seemed to disappear, presumably delivered back to the wrong section.

ARTHUR WALTER CLARKE

Government Analyst 1892-1893

Arthur Walter Clarke followed Mar and only served 6 months in the position. Clarke was born on 20 Apr 1854 in Cheltenham, Gloucestershire, England. He became a chemist, and in the 1881 census he appears as an assay chemist for the mines in Cornwall. Migrating to Australia on the Chyebassa, he arrived 19 Nov 1881; on is arrival in Brisbane, he was employed by Mr. B. Sparks, merchant, and subsequently by Messrs. A. Shaw and Co. He was next engaged by the Government to prepare a catalogue of the minerals that were sent as an exhibit to the Colonial and Indian Museum. This led to his being appointed mineralogical lecturer in North Queensland, which position he resigned two years afterwards and entered into partnership with Mr. Coane, of Charters Towers, as mining agents and analysts.

In January 1892 he was appointed Government Analyst in succession to Mr. R. Mar. Arthur Walter Clarke only held the position for around 6 months before he sadly took his own life. His death was reported in the Queensland Times, Ipswich Herald,

General Advertiser and The Argus (Melbourne), 2 June 1893. Even in those early days there appears to have been a lot of stress associated with workloads.

THE QUEENSLAND GOVERNMENT ANALYST.

SUPPOSED SUICIDE.

BRISBANE, THURSDAY.

The dead body of Mr. A. W. Clarke, Government analyst, has been found near Toowong. Death had apparently been caused by a shot from a revolver, which was found beside the body. The case was evidently one of suicide. Mr. Clarke was last seen at the Government Laboratory yesterday morning, when he was in good spirits, and he left, as he stated, to attend to some matters of business. Mr. Clarke was appointed Government analyst in January last, and was previously a member of the firm of Messrs. Coane and Clarke, analysts, Charters Towers.

EXPLOSIVES TESTING AND INSPECTION

The director of the laboratory was for a time also the Chief Inspector of Explosives. Samples were regularly tested throughout the 1970s. With the expansion of the mining industry in the 1970s and 1980s, this role was reassigned to the Mines Department.

TIN ORE BY FIRE ASSAY

A fire assay process was used to determine the tin content of tin ore. The process involved heating the tin ore in the presence of molten sodium cyanide. This reduced the tin oxide ore to tin metal which was then weighed. It was not uncommon to see the supervisor of the section boiling his morning tea on a bench surrounded by sodium cyanide crystals.

MYSTERY SAMPLE

A constituent of Senator Neville Bonner asked the Senator to have a sample of Galena (lead sulphide ore) analysed. The sample contained about 260 ounces of silver to the ton. There was no indication where the sample came from.

ANALYSIS BY FIRE ASSAY FOR GOLD AND SILVER

Fire assay involved melting a mixture of the ore with sodium *carbonate, borax, lead oxide, and flour (as a reducing agent) to form a molten glass. The process simultaneously formed molten lead which separated the gold and silver by extraction into the lead. The lead was then oxidised by heating in a furnace, in a special clay crucible (cupel), which absorbed the lead oxide and left a prill (bead) of the precious metal. This was weighed. Ambient temperatures reached 50 degrees Celsius in the furnace room. The colour of the residue in the clay cupel gave clues as to the presence of other elements. Care had to be taken to ensure the molds for the glass were dry, if not, molten glass would erupt from the molds and spray across the room.*

SHOW DOGS DIE FOLLOWING BATH

A member of the public rang to ask the lab to investigate why his show dogs died following a bath and shampoo. It turned out that the shampoo contained diazinon. It is relatively non-toxic. Unfortunately, the shampoo was old. Graham King found that the diazinon had broken down to a very poisonous compound tetraethyl dithiopyrophospate (S-S TEPP).

JOHN BROWNLIE HENDERSON FRACI

Government Analyst/Director

1883 - 1936

The appointment of John Brownlie Henderson in 1893 was the beginning of his 43 years of service as Government Analyst (Patrick, p. 135). The formal title GA disappeared in the 1950s but remained in common usage in some departments, especially police.

He had trained in chemistry in Glasgow, was always a strong advocate of one central laboratory servicing all departments instead of several small individual laboratories. With some specialised exceptions, this concept remains. With the increase in complexity of work, the cost of equipment and the increase in legislative requirements this is even more relevant today.

In his early reports, he pointed out that a large part of his time was occupied in examinations for the Police Department and, as a result, he was often absent from his laboratory, giving evidence before the various courts in the colony (Patrick, p.135). The introduction of telephone testimony in the 1990s and the acceptance of analyst's certificates by the courts reduced this burden.

Other departments submitted samples to the laboratory including hydraulic engineering which submitted artesian bore water for analysis. This service assured the safe use of water from the large number of artesian bores drilled in the early 20th century. The work continues to this time. It is interesting to note that a relative of the author was sacked from his job on a drill site because they struck oil not water.

Departments seeking his services included Mines and Geological Survey. The Port Master also sought his help in the determination of explosives. He gave evidence at the Royal Commission into the Mount Mulligan mine explosion which killed 75 workers. This laboratory role continued with other staff providing evidence to the investigation of the Box Flat Mine explosion in the early 1970s.

At first, very little of John Brownlie Henderson's time was employed in the examination of food and drugs. This led him to remark in his 1899 report that no attempt had been made by local authorities to enforce the provisions of the relevant legislation. Change came with the passing of the 1900 *Health Act*. In 1902 the analyst stated that the work of the newly established Health Department more than made up for the falling off in analyses for Customs, which became a Commonwealth responsibility on federation (Patrick, p. 135–6).

Local authorities also retained their powers. Some customs work was contracted back to the laboratory, particularly work which related to imports into Brisbane until the mid-1990s. At that time the Australian Government Analytical Laboratory, later the National Measurement Institute, were required to operate on a user-pays basis and withdrew the work, presumably to maximise their income.

John Brownlie Henderson commented that the new legislation was being ignored by local authorities with one exception; in 1903 the Brisbane Municipal Council employed a public analyst to examine food samples.

Five years later: The Commissioner of Public Health remarked that the officers of the council were cooperating in every way with the department to secure a pure milk supply for the city. The suburban and country councils had no interest in enforcing food legislation and appointed no analyst. Although the Commissioner of Public Health submitted a wide variety of foods for analysis, a major part of the laboratory's time was spent on examinations on behalf of the Health Department concerned with samples of fresh milk (Patrick, p. 136). In 1910 John Brownlie Henderson commented:

It is very unfortunate that the milk supply on which depends to a very great extent, our infant mortality should be in such an unhealthy condition. The purity of the milk supply is of very much greater importance from a health viewpoint than the purity of any other article of food, and yet it is the most adulterated and the most contaminated food material in use at the present time (Patrick, p. 136).

His report that year showed that water had been added to samples of milk in proportions varying from two to 32 per cent. The laboratory remained an essential part in the supply of safe Milk. This work continued until the deregulation of the industry.

THE TREE OF KNOWLEDGE

The Tree of Knowledge under which mythology has it the Australian Labor Party was formed was poisoned. The investigator called upon the lab to identify the poison used. After the tests were completed one of the staff, a lifelong supporter of the labor party, took the remaining leaves home.

MILK DEREGULATION

Although milk testing had been carried out for over 100 years, the work ceased when an agreement was struck with the milk industry to stop government testing under deregulation/ self- regulation of the industry. The laboratory was not informed of this change. Without a continuous supply of samples, it was difficult to maintain the equipment, NATA accreditation and competence necessary to support future legal prosecutions. The laboratory was directed to maintain competence. A small survey was carried out in the mid-1990s to assist in maintaining this competence requirement - at least one sample failed regulatory requirements. The laboratory was accused of violating the agreement. Milk testing was suspended.

GREEN FRUIT

In 1934 the Government analyst drew attention (The Morning Bulletin Rockhampton 22 Sep 1934) to a practise which has been recently adopted in respect to making green fruit appear ripe. "The green fruit is left overnight in gas charged rooms, and by morning a transformation has been made. The green has changed to yellow, and the fruit has most of the outward appearance of ripeness. But the appearance of ripeness is entirely fictitious, for the fruit within is still green and sour. Customers will not buy a second time from a dealer who sells them such worthless fruit, and the practise will hasten a growing tendency not to buy fruit from anybody. John Brownlie Henderson added, "The loss of market may help to stop this practise, but it is in the best of interests of the consumer and, therefore, also the

grower, that the problem created by the new fraud might be met with a new regulation deftly prohibiting such misleading artificial colouring."

YELLOW POWDER IN CEILING

In the 1970s a urea formaldehyde foam was sprayed into ceilings to act as a heat insulator. Over time this material breaks down to form a yellow powder. Many householders were concerned with the health aspect of having large quantities of dust in their ceilings.

Cairns Post 16 April 1937 (Connah's Appointment)

STATE ANALYST.

MR. CONNAH APPOINTED.

BRISBANE, April 15.

Mr. F. E. Connah, who has been Acting Government Analyst since Mr. J. B. Anderson retired in June, was appointed permanently to the position by the Executive Council to-day. Mr. L. A. Meston was made senior analyst.

Mr. Connah is a son of the late Mr. T. W. Connah, Under-Secretary to the Treasury and later Auditor-General.

FRANK CONNAH

Government Analyst/Director

1936 -1941

Frank Connah worked for the Queensland Government Chemical Laboratory from 1904 until retirement, being Government Analyst and Chief Inspector of Explosives from 1937. He qualified by examination for Fellowship of the Royal Institute of Chemistry. When appointed to the Government Analyst position the number of samples submitted by the Department of Health was approximately half of the total coming from all sources. In addition, a survey of bread and flour commenced, and attention was paid to the standard of drugs on the market, including the contents of headache powders and hair dyes. To meet the desire of the Department of Health and Home Affairs to improve public water supplies, the laboratory examined an increasing number of water samples collected by health inspectors throughout the state (Patrick, p.137).

LEON MESTON

Government Analyst/Director

1941-1946

Frank Connah retired in 1941 and was replaced by Leon Meston. Leon Meston joined

the Department in 1906, rising to Government Analyst and Chief Inspector of Explosives in 1941. He was educated at the University of Sydney (mining and science subjects 1898-99) and had wide experience in applied chemistry as assayer and metallurgist, cyanide chemist and researcher at mines in Queensland, New South Wales, and South Australia.
Courier Mail, 1 Sep 1941

He was the chemist in charge of manufacture and special investigations officer, Bureau of Central Sugar Mills, and water chemist with the Brisbane Water Board.

NEW ANALYST EXPERT IN TESTING "WATERED" MILK

Queensland's new chief Government Analyst and Chief Inspector of Explosives (Mr. L. A. Meston), whose appointment was gazetted last week, is the man who outlined a practically infallible method of determining the presence of added water in milk.

Mr. Meston read a paper on the method to the Royal Society of Queensland 30 years ago, not long after he had taken science and mining courses at Sydney University. Queensland adopted it then, the United States 15 years later, and England 27 years later.

Mr. Meston is a son of the late Mr. Archibald Meston, who was an authority on aborigines.

By 1941 the laboratory was now sectionalised, with one section dealing solely with the analysis of food and drugs. Examination of milk continued to consume a large part of the section's time, and that perennial bone of contention between the Health Department and butchers—the prohibition on sulphur dioxide in mincemeat produced work for the analyst. Meston's reports show that in many samples of paint the amount of lead was higher than the five percent permitted by the current legislation (Patrick, p. 137). Lead in paint and sulphur dioxide in mince meats continued to be tested in the laboratory up until at least 2012. Meston retired at the end of 1946, after 40 years of service.

SILVER 50 CENT COIN INGOT

In the late 1960s decimal currency was introduced into Australia. The new 50 cent coin was a round coin with a high content of silver. Over time, the silver content became more valuable than the face value of the coin. Criminals collected these coins and attempted to melt them down and sell the silver. The laboratory was involved in providing confirmation that the ingots came from 50 cent coins.

EXHIBITS TAMPERED WITH

The laboratory routinely destroyed drug exhibits for the police. The analyst who was on duty recognised an exhibit being returned for destruction. The original sample he had analysed was a pink rock heroin. The sample was now a white powder. This demonstrated the advantages of staff who are observant and carry out all stages of the process. This pickup would not occur when dedicated staff are responsible for the destruction, as in a production line approach.

INTERESTING CALLOUT

Sometimes you get to go to places you would not normally be permitted. A request was received from the Fire services to provide a chemist to give advice at the airport. In the main luggage compartment of a commercial passenger jet a bottle of liquid had broken and spilled. This was a formaldehyde solution used to preserve the gonads of sheep. The bottle had broken in transit and the liquid had leaked through the floor seams into the underbelly of the aircraft. The author spent the night in the engineering workshop at the airport whilst the aircraft was dismantled to clean the various parts. An interesting experience.

COAL DUST IN COAL MINES

In the early 1970s, the Mining Inspectorate decided to undertake a survey of stone dusting compliance in coal mines. Limestone was required to be spread on the coal dust to reduce its combustibility and hence the probability of an explosion. Laboratory staff took samples from the Box Flat Coal Mine on Thursday and were due to take more on the Tuesday. On Monday, it exploded, killing all 17 miners on the shift. This laboratory provided evidence into the investigation of the Box Flat Mine explosion. The disaster led to a survey of all the coal mines in the State. If a mine was found to be deficient in its stone dusting, operations were suspended

until they prove they had complied with the regulations.

BOOKING AIR FARES

Budgets were always tight, so staff shopped around to get the cheapest airfare for interstate and overseas travel to meetings and conferences. The department developed the concept of the travel hub which booked all airfares for travel by staff. There was no

option to book your own ticket. The cheapest ticket was seldom purchased. The dearer ticket was charged against your budget. Approval of overseas travel was by the minister. On one occasion a staff member only received approval a few days before departure resulting in very expensive fares. It can be frustrating working in Government.

STEWART BYRON WATKINS

Government Analyst/Director 1946-1960

Meston was followed by Stewart Byron Watkins who completed his Master of Science in 1918 and lectured in chemistry at the position of Supervisor of the Central Technical College, from 1919 to 1941. He travelled throughout Queensland from 1941, when he worked with the toxicological section. He published many papers and articles on horticulture and was appointed to the Council of the Horticultural Society of Queensland. Stewart Byron Watkin expressed his philosophy relating to the tasks of the analyst in the following 1948-49 report:

A large proportion of the activities of the laboratory, undertaking as it does a preponderance of chemical assessments, must of necessity be routine; however, analysts should not be regarded as automatons they are officers trained to an appreciation of their professional responsibility and able to exercise initiative in various problems which do arise even in the case of what is often referred to as soul destroying routine (Patrick, p. 138).

Watkins pointed out that his staff were called upon to consult on stream pollution, location of industrial plants, fluoridation of water supplies, production of oil from coal and the establishment of food and drug standards.

His reports of the time reflected on the progress that had been made in food processing, and the advent of new organic chemicals after the Second World War. His analysts were looking for fungicides in paint and DDT and other new chemical substances on vegetables and fruit. In his last report before retiring in 1960, Watkins wrote that the standard of the milk supply was easily the best on record due in no small measure to the increase in consumption of bulk processed milk. He pointed out that while paint scrapings from old buildings still contain lead in quantity, none of the new household paint contained lead (Patrick, p. 138).

MERCURY IN HOSPITALS

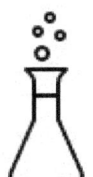

As part of our contract monitoring role, the laboratory assessed the quality of mercury thermometers used in Queensland Government hospitals. Each thermometer contained about 0.5 grams of mercury. The author asked how many thermometers were used yearly. The answer was about 48,000. Next question, what happened to the thermometers, are they only used once. The answer was that they are thrown away when they break. This means that potentially 24kg of mercury was splashed around hospitals when the thermometer breaks, depending of course where they break. The broken ones were thrown in the wastepaper baskets. Initially when approached, the department claimed that the patient bit the thermometers and swallowed the mercury. A hospital worker stated that this was nothing compared with the weight of mercury spilled when mercury blood pressure gauges were dropped.

The laboratory took a mercury vapour meter to a hospital and tested the contents of the vacuum cleaner floor polishers. The meter went off scale. Sometime later, another Government department put an order on one of the hospitals to fix the problem when their air measurements showed high levels of mercury in the ward air. Perhaps, the laboratory's actions facilitated the introduction of electronic thermometers and electronic blood pressure gauges, eliminating the problem. Many of the older hospitals have been rebuilt since that time so the problem of residual mercury may not be an issue. With time, mercury droplets develop a dirt and oxide layer which reduces the vapour level in the surrounding air.

KETAMINE ABUSE 1980'S

A syringe was received from the police which upon analysis was found to contain ketamine. At the time Ketamine was a veterinary anaesthetic. The laboratory had not previously encountered this substance as a drug of abuse. Around the same time in another section of the laboratory, a urine sample was received and analysed. It also contained traces of ketamine. This result was published in a forensic journal. Ketamine became a drug of abuse.

IS IT FROM SPACE?

A member of the public rang a local radio station to report a large amount of metal had appeared on their driveway. The metal had been molten at some stage. The radio station put out a call for a chemist to investigate. The GCL responded.

The usual precautionary radiation checks were made. There were no signs of burned ground, so we were satisfied that it had solidified elsewhere. Analysis by XRF showed it was a common alloy.

CHAPTER 3: DIRECTORS AND SENIOR MANAGERS - 1960–2023

IVO HENDERSON
Director 1960-1973

Ivo Henderson followed Byron Watkins in the position. He was a Chemist, and an internal appointment; he had worked in the GCL for many years. During his time a scholarship system was devised which recruited students with high academic achievements in chemistry at senior or first year level. The scholarship paid their university fees and a living allowance but included a four-year bond to work at the laboratory. It also provided work for the students during the Christmas vacation. Two scholarships were awarded each year, a total of eight scholarships. All scholarship holders except one who left the organisation to become a teacher, rose to high management levels within the organisation. The author was one of these. The concept of recruiting the best students rather than waiting for them to graduate, and then competing with other laboratories to attract them, demonstrated Ivo Henderson's strategic vision.

This period marked the beginning of the end of the wet chemistry era. Spectrophotometry and titrations were used for the bulk of the work. Instruments were gradually being introduced into the laboratory, including the first gas chromatographs (GC).

The first Gas Chromatograph (GC) in the laboratories was built by one of the staff, Garth Lahey. He utilised a kerosene can of boiling water as a constant temperature environment in place of the oven. The staff routinely packed and conditioned their own columns. The first atomic absorption spectrometer (AAS), a Varian Techtron Serial number 0033, was placed in the Waters section. It was in use for the next 25 years.

The concept of an 'instrument hub' had become the benchmark of efficient analysis; however, this was not always the case. This original AAS within the GCL was installed in the Paint Testing lab, after a later model was purchased for the mining area. It provided results within minutes of the samples being digested and diluted, while submission to a separate laboratory hub sometimes took days. Industry learned this lesson and replaced the central analytical facility with instrument on or next to, the production line. Of course, with complex, expensive specialised instruments, central instrument hubs are essential. It is interesting to note that when an instrument had mechanical knobs and dials, most analysts were able to learn quickly how to use them. Once the dials and knobs were replaced with computers, the operation of the instrument became a more specialised task.

In 1962 Ivo Henderson was a member of a new committee with the aim of maintaining control of the Laboratories in the Public Service.

The 1960s were a period of significant staff turnover. The mining boom resulted in the loss of many staff. In order to attract staff back to the laboratory, the wages of chemists were doubled, which resulted in an influx of young inexperienced chemists, the majority of whom were less than 24 years old in 1972. Given the narrow age difference, we once joked that in 40 years' time we could all retire at once and turn the lights out. Little did we know that with the mass redundancies of 2012, this was almost the case. The first female chemists started working at the laboratory during the 1960s.

Ivo Henderson was more outwardly focused on his activities and left the day-to-today management of the laboratory to his deputy director. Ivo Henderson retired in 1973.

FIRST AND SECOND WORLD WARS

In the First World War, a number of members of staff served in the military. During the Second World War chemists were not subject to military service as they were regarded to be of strategic importance and exempted. They carried out analyses vital to the war effort. There was also interest in

claims concerning the levels of vitamins in various foods on the market. In the latter years of World War Two, there was an increase in the amount of food analyses performed by the laboratory. The Commonwealth Food Control Organisation, which was established to provide food to American and Australian forces, made large demands on laboratory services (Patrick, p. 134).

LEUKAEMIA CLUSTER

In the 1970s, air testing was carried out at Emerald to determine exposure of local children to pesticide spray drift from cotton farming.

PAINT LIFTING FROM FLOOR

The lab received a request complaint from a builder that a floor paint job had failed. George Lee was assigned to investigate. He noted that the failure was of a geometric design in rectangles. He asked what the building had previously been used for and was told it was a soft drink factory. In the past soft drink bottles were recycled after washing, usually with a highly alkaline solution usually sodium

hydroxide. George checked the pH of the floor, and it was still highly alkaline. This explained the failure of the paint job. Without a broad generalist knowledge of industrial processes, this problem may not have been so easily resolved.

ILLEGAL TRANSPORT OF DANGEROUS GOODS

A reference material was required, and a USA based company identified on the internet as a supplier. The order was through Hong Kong. When the chemical arrived, it had a different name on the label. An email was received from the supplier stating that for transportation reasons, a false name had been supplied with the chemical and that it should be relabelled with the correct name. The bottle was also leaking. It was amyl nitrite. This was a highly dangerous practice and was reported to the authorities. In the past when an alkyl nitrite chemical was required it was synthesised in house. One wonders if the training of modern chemists still involves the same level of chemical synthesis training as occurred in the past. Synthesis is a legitimate route to the preparation of chemical reference materials.

Proceedings of the Royal Society of Queensland 1955 reported:

The reports of the Government Analyst make fascinating reading if you allow for the cold, official phraseology. The following was taken from the first (1882), and the (1954) reports.

Hopeful prospectors are still apparently looking for gold in mines near Brisbane because Mr S. B. Watkins states in his last rep: A metal submitted as a mineral from the Mount Coot-Tha area consisted of brazing alloy. It carried a patina, and no doubt was a relic of the occupation of the area by the American forces during the war.

DAVID MATHERS
Director 1974-1978

David Mathers followed Ivo Henderson as Director of the GCL. He was a Chemist and previously the Deputy Director of the laboratory. During his career he had risen through the various positions to become the last person to bear the title Director who was officially appointed from within the laboratory. He was a great supporter of the concept of managing by walking around, long before it became a management catchword. He was commonly seen several times a day in each of the laboratories, often correcting the techniques of young chemists.

As the Government Analyst, he reviewed and signed every report that left the laboratory. His task was made more difficult by the relatively inexperienced young chemists. He was the last Government Analyst to have the role of Chief Inspector of Explosives. With the growth of the mining industry, the Mines Department took over this role. David Mathers managed an extensive cadetship program to recruit technicians. Students with good passes in senior chemistry were hired on the basis that they would do a Certificate in Chemistry part-time, at night, at what is now Queensland University of Technology (QUT). He lectured at QUT and was on their Chemistry Advisory Committee. The cadet program set the laboratory up with a cohort of very competent technicians, many of whom served their entire careers at the laboratory. Some were discouraged from doing a full degree which limited their career prospects until the salary scales were combined in 2010. The work was dominated by food, forensic drug, toxicological, water, customs and mineral analyses and work for the State Stores Board to assess contracts for textiles and cleaning products. Staff transfers between sections were routine and usually done without consulting the affected staff member. Transfers often occurred when passing the director. For example, the author was told, while walking down the corridor, "when you get back from holidays start work in Toxicology."

For more detailed information on the period following Dave Mathers departure in 1977 and Trevor Beckmann's appointment in 1981, see Ron Biltoft's personal story Chapter 25. This was a period of appeals against appointments, restructuring, rivalry, consultants, and long-term acting positions.

EXPLOSIVES ANALYSIS

Gelignite was frequently tested for stability. It was common for laboratory staff to collect a large box of gelignite from Roma Street, which was then driven up George Street to the laboratory. Nitroglycerin causes a drop in blood pressure, and many staff suffered headaches as a result of performing this test. An explosive magazine was situated on site in William Street and was often filled with gelignite samples sent in for stability testing. Periodically Christmas crackers were received to determine the weight of explosive material used. Various ammonium nitrate samples were tested for use in the production of ammonium nitrate fuel oil explosives. The anti-caking agents used to form the prills potentially affected the explosive potential.

TREVOR BECKMANN
Director 1979-1988

Trevor Beckmann was a Chemist and was appointed from outside the laboratory. He was previously the Deputy Director of the Queensland Agricultural Chemistry Laboratory which was part of the Department of Primary Industries (DPI). That laboratory was located immediately adjacent to the GCL on William Street. In the mid-1970s, the DPI laboratory had successfully gained funding for new facilities, and they moved to a new site at Yeerongpilly. This move allowed an expansion of the GCL into the laboratory previously occupied by the DPI, effectively doubling the floor space. Given the deteriorating state of the William Street laboratory, Trevor Beckmann made it a priority to gain funding to build a new laboratory building. In 1983, he was successful in obtaining funding for a new laboratory included in the government's Capital Works program. It helped that he personally knew the Health Minister who he went to school with. Over the next five years, much of the laboratory's discretionary time was spent designing the new facilities. Don Lecky was assigned to coordinate the design.

I am led to believe that Trevor won a scholarship for his Masters. He did it in France and was assigned the same desk that Madam Curie used at one stage. How's that!

There had been very few promotions to senior chemist positions at the laboratory. Trevor Beckmann introduced a set of criteria which had to be met to be eligible for promotion. Criteria included published research papers, as well as a thesis on some aspect of the work performed by the applicant.

The laboratory undertook work for the Commonwealth Customs Department on a fee-for-service basis. The Queensland Government was keen to ensure that they were charged the correct amount for the service. A departmental review of the cost of all work performed by the GCL was undertaken. After this review, the laboratory was required by the Treasury to implement a user-pays system, however, it managed to maintain its budget funding for traditional clients like the Health Department and the police. The new facilities that Trevor Beckmann was instrumental in bringing about were completed in 1989 just after his retirement.

 ## AFLATOXINS IN PEANUTS

Many analyses have political implications. The laboratory routinely performed the determination of aflatoxins in peanuts.

During the 1970s, a shipment of peanuts from overseas was destroyed due to it exceeding the limit for aflatoxins. However, I heard that when the Queensland crop showed signs of aflatoxins the permitted limit was increased to enable them to be sold.

MUSTARD GAS - THINGS ARE NOT ALWAYS WHAT THEY SEEM.....

A large amount of mustard gas was stored in Queensland during the Second World War. After the war, some of this chemical warfare agent was buried and other quantities dumped at sea. The laboratory received a request from a farmer who had found a military item which had a liquid inside. There was the possibility that this may have been a chemical warfare agent, and the laboratory was not properly equipped to open the container. The Defence Department became involved, and a team opened the object and found it was full of creosote, a wood preservative.

The object was an artillery shell container which a farmer at some stage in the past had used to store creosote. Another sample marked mustard gas was received which upon infrared analysis turned out to be CS teargas.

... .SOMETIMES IT IS MUSTARD GAS

Trevor Beckmann also enjoyed working with the fire brigade on chemical incidents. On one occasion, he was called to Moreton Island to investigate an object on the beach which was originally thought to be a bomb. When it was detonated a brown liquid exuded from the object. This liquid turned out to be mustard gas. It was taken to Helidon where it was treated by two chemists and buried. It was one of the tons of chemical munitions dumped in the sea off Moreton Island after the Second World War. These still occasionally wash up on the beach.

Mustard gas (liquid) tends to degrade to form a thick outer layer in sea water which encapsulates the still active mustard gas liquid within. Mustard gas shells are still periodically uncovered in Queensland.

DES CONNELL FRACI

Director 1990-1994

Professor Des Connell came to the laboratory in 1990 from Griffith

University and took over management of the combined laboratory at the new facility at Coopers Plains. On his first day, he drove his car into the director's car space only to be approached and directed to move his car by the security guards. He had previously worked in the Department of Primary Industry laboratories. A hundred year previously Mar's duties had not been well defined by the government; when Professor Connell asked the department what they wanted the laboratory to do, the answer was 'be there when we need you'. This statement encapsulated the role of the laboratory. It is to carry out any urgent analyses for emergency responses, public outcry, and media concerns as required by the government. On numerous occasions, results from the laboratory have been used to reduce public stress and media concern.

The laboratory was involved in numerous emergency responses over the years (Chapter 21). Such matters were always given priority. To meet this requirement, it was necessary for the staff to be competent in carrying out a wide range of analyses. This competency could only be maintained through routine analysis. Systems had to be operational at the time of the emergency, as there was no time to develop and validate new methods. Nonetheless, the laboratory had the capability of responding to any given emergency at any time, often completing analyses in a matter of hours. Due to the presence of a wide range of systems, the analysis could be performed in a matter of hours on a 24-hours-a-day, seven-days-a-week basis, often before a private lab could provide a quote to carry out the work. The GCL never failed to respond to government emergencies with data to enable decision making.

During this period, consideration was given to the formation of a Queensland Health Science Institute (QHSI), and a considerable amount of time was spent developing a structure for this organisation.

Professor Connell initiated actions to amalgamate the Laboratory of Microbiology and Pathology (LMP) with the GCL. This amalgamation was the first step in the growth of the laboratory into new areas of science. The LMP was situated in George Street, Brisbane. The building had reached the end of its useful life and was not fit for purpose. Further funding was obtained for the construction in 1998, of a new Virology laboratory with PC 4 containment facilities and to renovate part of the building previously occupied by the Environmental and Potable Water sections of the Chemical laboratory into a Microbiology laboratory which would house staff from LMP.

Des Connell had a research background and had a role in stimulating research activities among the chemistry staff. Carrying out long-term research in a high throughput, short turnaround time, routine chemical analysis laboratory environment proved to be difficult. In research laboratories, it is common to see equipment worth millions of dollars sitting unused and thus available. In an operational, production-based laboratory, emphasis is placed upon the maximum utilisation of equipment. With a regular flow of samples with short turnaround times, it is difficult to pull an instrument offline for weeks to answer a research question.

The Director-General of Queensland Health at the time, Dr Peter Livingston, became aware that the National Health and Medical Research Council (NHMRC) was preparing to advertise an organisation to host a new specially funded research centre on

environmental health. He requested Des Connell to coordinate with Professor Susan Pond, a prominent researcher, in the Department of Medicine at the University of Queensland to prepare a submission. The outcome was successful, and Queensland Health was awarded an NHMRC grant for a research facility into environmental toxicology.

It was anticipated that this new unit would work in conjunction with the GCL, and Des Connell was instrumental in providing space within the laboratory for the National Research Centre for Environmental Toxicology (NRCET).

The creation of NRCET helped overcome some of the difficulties in carrying out research in a routine laboratory. NRCET researchers were at the cutting edge of environmental toxicology, they utilised GCL equipment as well as their own to carry out research. This decision had long term positive consequences for the laboratory. Des Connell commenced the laboratory on the road to obtain ISO 9001 quality system certification to supplement its previous NATA accreditation. In the mid-1990s, ISO 9001 quality systems certification was obtained. This certification became a requirement for companies seeking government contracts, including government laboratories.

An unexpected outcome of ISO 9001 was that government departments accepted certificates from accredited suppliers and decided that it was not necessary to test the product against specifications. For example, paints used on government buildings and textiles used by government agencies ceased to be submitted. After many years of providing this service the sample flow dried up, and the laboratory suspended this service and moved into other areas.

During the drought in the late 2000s, recycling of water became a major political issue. Fortunately, the chemistry laboratories were already committed to water testing, carrying out analyses of a large range of pesticides, drugs, pharmaceuticals, and other pollutants. The laboratory, and therefore the Queensland Government, had an extensive range of analytes because of the collaborative research work carried out with NRCET and the organic chemistry section under Mary Hodge's direction.

The end of the NHMRC funding for the centre resulted in it changing its name to Environmental Toxicology (ENTOX) and continuing to be co-located within the laboratory. It remained there until 2018 when it moved to another location. Des Connell returned to academic life at the end of his five-year contract.

TURNAROUND TIME COMMITTEE

Over its long history, the laboratory has been criticised for poor turnaround times. In 1993 Geoff Rynja, Willie Gore, Lenore Hadley and Gary Golding with Des Connell's support formed a committee with an aim to reduce the turnaround time on samples throughout the laboratory. Strategies were developed and implemented. The turnaround time in Investigative Chemistry for Health work changed from the oldest sample being six months old to the oldest being three weeks old. When this was brought to the attention of one of the clients his reaction was 'you mustn't have much work to do.' The acronym TAT 'turnaround time' entered the language of the laboratory as a performance indicator.

One technique involved concentrating on getting rid of the oldest sample. This technique was taken straight out of the book by Dale Carnegie, How to Win Friends and Influence People. The Supervisor, Gary Golding, wrote a number on the door of his office, which corresponded to the age of the oldest sample in days. Staff became curious about what this number meant. When it was explained to them, the supervisor found the next day the number was crossed out and a smaller number substituted. This continued until satisfactory turnaround times were achieved.

OIL SPILLS FROM SHIPS

During the 1970s to 2000s, the laboratory was involved in the comparison of oil spills, with oils from various boats that were present in the area at the time of the spill. The objective was to 'fingerprint' the oil so that prosecution could be undertaken.

DON LECKY
Director 1989-1995

Don Lecky commenced work in the laboratory in 1946, having worked his way up through the structure to manage the day-to-day activities of the GCL during Des Connell's time as Director. Following Des Connell's departure, Don took over the director's role before his retirement. He worked extensively in the forensic areas and developed excellent relationships with the Queensland Police Service (QPS).

During the 1980s, Don Lecky was given the task of coordinating the design of the new facilities at Coopers Plains. The laboratory was completed in 1989.

Under his management, the laboratory balanced its budget—managing the chemistry laboratory during the recession of 1993 presented some severe

challenges. Budgets were not increased, there was a loss of approximately 10 staff positions, and instrument replacement was not funded. Fortunately, the laboratory was able to utilise earned funds to fill gaps.

The shortage of funding within the government effectively ended with the introduction of a Goods and Services Tax. This taxation change led to the widespread replacement of equipment and the updating of technologies with the introduction of a capital equipment replacement program within the department.

Don Lecky also developed a collaborative agreement with the Department of Emergency Services to supply five chemists to attend chemical emergencies and to advise the fire services on the safest and most efficient method of resolving a chemical emergency. This collaboration is still being carried out in 2023.

Don Lecky retired after 43 years of service with the laboratory.

PAUL GEOGHEGAN
Director 1995-1998

With the departure of Don Lecky, Paul Geoghegan took over as Director of the GCL. This period was a time of considerable change in the laboratory. Decisions at higher levels of the department had been made to amalgamate the chemistry laboratories; the microbiology laboratories and the hospital pathology laboratories under one general manager. The mortuary and its associated services had already co-located to the Coopers Plains site as stage two of the scientific complex in the purpose-built mortuary named in honour of Dr John Tonge.

Before this time, the GCL had been autonomous and able to make its own decisions. Decisions were often made and quickly implemented. Becoming part of a large organisation changed the political environment; and it marked a change from a small business structure to a corporate structure. Internal politics and the management of these became critical.

As part of this process of amalgamation, one morning, the entity known as the GCL ceased to exist and was replaced by the amalgamated Queensland Health Scientific Services (QHSS). There was no ceremony to mark the end of this century old organisation or one to welcome in the new. After overseeing stage three construction at Coopers Plains to accommodate the tertiary work of the Laboratory of Microbiology and Pathology, Paul Geoghegan retired after 30 years of service to the laboratory.

MURDER PLAN

A citizen approached the police to inform them that a person had approached them to obtain some sodium cyanide. The police set a trap. They asked the laboratory to make up a block of material that looked like sodium cyanide. They made this using a mix of sodium chloride and potassium chloride, pressed into a block using an infrared press. Some days later the analyst received a call saying the person seeking the potassium cyanide was found dead. He was concerned that his mock cyanide had killed him. After letting him worry for a while, he was told that the person had shot himself. He had presumably tried to suicide using the mock cyanide and it did not work so he used an alternative means.

CANCER CLUSTER ABC STUDIO TOOWONG

The Laboratory carried out a range of chemical and radiological tests.

MICHAEL MOORE
Director 1995-2004

The amalgamation of the different laboratories into a much larger organisation led to organisational restructuring. Professor Michael Moore was appointed the Director of the combined QHSS laboratory. He was a Scotsman from Glasgow, like John Brownlie Henderson some 100 years before. Professor Moore was a world authority on environmental toxicology and a member of many national and international committees. As well as being the Director of QHSS, which incorporated the GCL, and the Laboratory of Microbiology and Pathology, he was also Director of the National Research Centre for Environmental Toxicology (NRCET) co-located in the building. This dual role led to some perceptions of conflict of interest among staff in competition for resources and fee-for-service income. There was also some conflict between the bureaucratic approach of Queensland Health and the more laid-back approach of a research agency, especially around public statements. Michael's role was to turn the laboratory into a research establishment. A considerable number of students from NRCET worked in collaboration with laboratory staff to improve and increase the capability of the labs.

This collaboration mainly focused on the Organics section and was led by Mary Hodge, the Chief Chemist of the group. This collaboration resulted in a considerable number of papers being published and the laboratory moving to the cutting edge of environmental chemistry. The rise of specialist scientific groups changed the relationship between the laboratory and those seeking chemical advice. In the past, the GCL contained some of the only technical chemical experts present within the government.

They were often called upon for a wide range of advice. In many respects they were generalists; one day they may be giving advice on contracts for laundry chemicals, the next day toy safety, and the next on chemical warfare agents. Despite being regularly required to provide advice on a broad range of areas, the advice was always sufficient to satisfy the needs of the government.

The Internet also arrived in the early 1990s and was made available within government offices in the late 1990s. The Internet provided clients with a means of accessing information directly themselves. This access reduced the role of the laboratory in advisory matters. Somebody elsewhere in the world had usually investigated possible public health problems, and this information was readily available and could be used in the Queensland context. This expanded knowledge base reduced the number of samples received in some sections and necessitated the development of other areas of work. Health

Inspectors who had previously called upon the laboratory for advice checked the Internet first. Retitled Environmental Health Officers, they were now tertiary qualified and able to do their own research.

Before the establishment of QHSS, the Laboratory of Microbiology and Pathology had divested its clinical work into the hospital pathology laboratories. While this divestment freed up staff to perform research, it primarily remained a tertiary referral laboratory. Some chemistry laboratories were limited in their ability to divest their work. Some work was phased out, such as the geological survey work, which commenced in the 1930s. The downside was that in divesting this work, the laboratory was required to divest the budget which had been allocated to this work. **Drawing courtesy of George Lee - Later drawings featured a small outboard motor.**

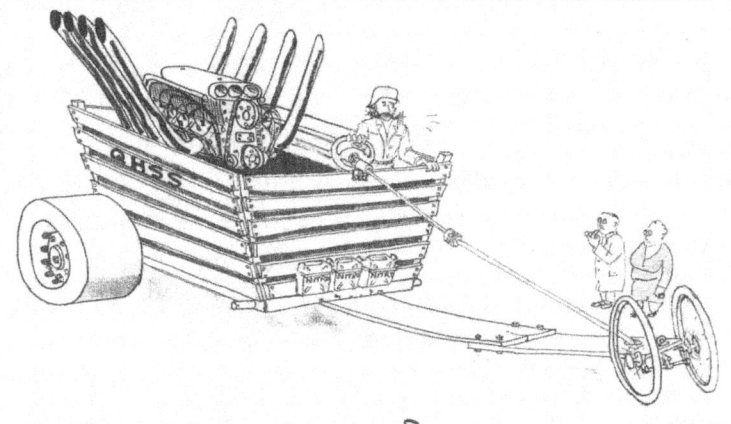

WANNA WEE RIDE ?

This meant that no financial resources were freed up for high-level research work. This quandary continued to inhibit the move to new types of work.

Although the laboratory continued to be present on many government committees, over time the government started to call upon NRCET for interpretive advice on environmental toxicological matters. This advice had previously been provided by the chemists of the GCL. The laboratory's role supplemented this advice, especially around analytical methodology.

At one point, Professor Moore proposed the introduction of short-term rolling contracts for all staff. He had worked in this environment in a previous organisation. When asked for a sample of the contract, he was unable to provide one, as all the staff he knew at the previous organisation were no longer employed there. Staff breathed a sigh of relief when this concept ceased to be a priority. During this period, the laboratory had its first visit from a series of management consultants whose aim was to increase efficiency—usually via restructuring and formal business planning. This resulted in a reduction in the number of chemistry sections and removal of middle management positions. On one occasion, a new structure was circulated where whole sections had been deleted without any consultation. Over time, the five senior chemistry management positions of the early 1990s became one position. This change reduced time for strategic activities and personal interaction between senior management and the growing number of staff on site. The concept of the organic and inorganic chemistry sections was created. This structure represented a focus upon analytical process rather than outcomes for clients.

One of Michael's lasting legacies was the construction of a large conference room facility which has hosted many conferences and lectures over the years. This raised the research and public profile of the laboratory.

GLASS IN CANNED FISH?

Glass-like fragments of struvite (Magnesium Ammonium Phosphate) are occasionally found in canned fish.

ANAESTHETIC GAS IN OPERATING THEATRES

In the early 1970s, for a period, the anaesthetic gas Halothane was used in operating theatres. A number of staff who worked in the operating theatre experienced miscarriages related to the gas. Later, expired air was passed through a carbon filter to reduce the concentration of the gas in the operating theatre air. A portable GC was used to monitor levels.

BACK-LOGS OF SAMPLES

In the past 50 years there have been consistent complaints about the long turnaround times on sample analysis. Most samples just turn up at the lab unannounced. Where the samples are sent on a particular day by appointment, the turnaround time drops to minutes or days not weeks. Microbiologists solved the problem, out of necessity. Delays in analysis can lead to growth of bacteria. As such they required appointments for submission of microbiological investigative samples.

LEAD IN PAINT

Analysis of lead in paint continued until about 2022. Levels as high as 30 per cent were occasionally found in paint scrapings from old Queenslander houses. Despite almost 100 years of effort, the hazard remained. During renovations on the First World War German tank 'Mephisto', the Queensland Museum asked that samples of paint be tested for lead content. Staff enjoyed the opportunity to climb inside Mephisto. The samples contained lead as would be expected from paint of that vintage. The radiator of the engine was inside the crew cabin. Nice in winter.

RON BILTOFT
Centre Manager Forensic Sciences 1995-1998

As part of the restructuring to form QHSS, the laboratory was divided into two sections based on 'forensic' and 'public health' sciences. It had been structured on a discipline basis for the previous 125 years. This new structure meant that the forensic chemistry laboratories, the forensic biology laboratories, and the mortuary pathology services were combined under one manager. The public health chemistry laboratories and the microbiology and virology laboratories were also combined under another manager. The managers were thus, not discipline experts in all of the areas they managed. However, Ron Biltoft, through years of generalist training as an analyst covering a wide variety of requests, quickly gained sufficient expertise to manage most situations.

During the late 1990s, resourcing of the laboratory failed to keep up with demand due to the recession and uncompleted work began to grow. For the next ten years, considerable effort was put into obtaining additional staff to carry out the work.

Clandestine laboratories became a significant source of political and media interest. Due to the accumulated work, court dates would sometimes be missed, and there was criticism from the judiciary and media. This could mean that the defendants were released on bail and committed further drug manufacture offences which would not have occurred if they had been jailed promptly for the first offence. The accumulated work continued to grow, and samples had to be analysed out of order due to the higher priority of some cases. This delayed other cases.

Media criticism was not always justified. Over many decades it was noted by staff that prosecutors or police claimed that the analyst had not completed the analysis solely to get an adjournment. On occasions, they were telling the truth as the sample had only been delivered to the laboratory the day before the case. Defence lawyers were often upset that their clients had languished in jail for weeks awaiting the analytical results when the samples had not been submitted. This delay in the submission was in part the result of the police waiting to see if the suspect pleaded guilty before submitting the samples. If every sample was submitted upon arrest, the laboratory's workload would have increased threefold.

Doubling the drug squad was a common election promise, nobody thought of the consequences for the laboratory. Ron Biltoft was instrumental in setting up a high-level liaison group which included the Queensland Police Commissioner. This group aimed to overcome past problems and to lobby the Government to provide additional resources for the forensic areas. This action and the efforts of others resulted in a doubling of the number of staff in the forensic areas over the next 10 years. Ron Biltoft retired in 1999 after 36 years of service.

ABRUS PRECATORIUS {ROSARY PEA} NECKLACE

In the mid-1970s, Chemist Ron Biltoft saw some red and black seeds on necklaces in passing a shop in George Street, Brisbane. He recognised a potential hazard. In 1974 his family had gone on a cruise around the Pacific, and bought some jewelry made from the same seeds. When he got back to Australia, quarantine confiscated it and told him that the seeds were extremely poisonous. They were Abrus Precatorius (rosary pea) that contained the toxic substance abrin. Less than three micrograms of the abrin are sufficient to kill. People drilling holes in these seeds have been poisoned by the material adhering to their fingers. Even today's technology is unlikely to pick up this toxin unless chemists were specifically looking for it. Ron contacted the Government Botanist, Harry Kleinschmidt, and worked out a way to denature the poison by heating the seed before manufacture of the necklaces. He took all the information to the Chief Inspector of Drugs and Poisons to see if the seeds could be heated before being made into the jewelry. The Chief Inspector said he'd 'look into it'. Perhaps he had a quiet

word with the shop. The jewelry remained on sale, hopefully heat treated.

LENORE R

Centre Manager Forensic Sciences 1999-2002

Could not be contacted to obtain permission to use name and image.

Lenore R replaced Ron Biltoft as Manager of the forensic sections. She had previously worked in a pathology laboratory in Cairns. This was a period of lack of resources and accumulating sample backlogs. Balancing the budget became the prime performance indicator and thus a priority for management. The three rules were, don't go broke, don't go broke and don't go broke. In 2002, the first Liquid Chromatograph Tandem Mass Spectrometer was purchased for toxicology. Constant media attention concerning forensic turnaround times led to an interdepartmental enquiry into forensic science.

ANNUAL HAPPENINGS

Each year the lab receives complaints from the public that there is a residue floating in their water tank. On occasions the white material was identified as LERPS. A lerp is a white sugary waxy covering of the immature stages of psyllid insects (Family Psyllidae). On other occasions the substance was pollen from gum or pine trees.

ALEXANDER (SANDY) F

Centre Manager Public Health Sciences 1995-2006

Could not be contacted to obtain permission to use name and image.

Sandy F was a Scotsman with an extensive background in laboratories having managed the Laboratory of Microbiology and Pathology. He had divested the clinical work from the LMP to the hospital pathology laboratory system. This divestment freed up some resources for research. Public health was promoted as the purpose of the laboratory, and the Health Department was, thus, the client to be satisfied.

This strategy did not fit well with the chemistry laboratories, which had a diverse range of historical non-health and commercial clients. Work for the health department in the area of food testing had diminished over the years, drugs and poisons had been taken over by the Commonwealth. Most health samples were related to a specific problem rather than routine surveillance. The director general at the time is reported to have said about the work of the chemistry laboratories, "this is all fine but what has it to do with health". The high-level management seemed to have missed the fact that the Health Department had been given custodianship of the laboratory because it had to sit somewhere within Government but did not necessarily do work exclusively for that Department as occurs with the hospital pathology laboratories and the Laboratory of Microbiology and Pathology (LMP). Also, public Health is the fence at the top of the cliff that avoids the need for an ambulance at the bottom.

The chemistry laboratories were in a difficult position in that if they divested work, they would have to transfer an equivalent budget to the new provider. If they ceased their commercial work the laboratory, as a whole, would fail to balance its budget. During this time, many laboratories in other states and overseas were being commercialised, closed, or put on a full fee-for-service basis. Ceasing the commercial work for the government and non-government clients would have placed the laboratory at a severe disadvantage should a similar decision be taken in Queensland.

The blending of the cultures between LMP and GCL staff proved to be challenging. In the Microbiology laboratory the clinical client (a doctor) usually decided the appropriate test as a result of a diagnostic examination and a pathologist signed the report. In the Chemistry area, when the client had a problem, the chemist had to decide on the test to solve the problem. For example, the substance may be an unknown chemical, and the analyst had to decide how to identify it. Microbiologists never wore their lab coats outside the lab for good reasons. Chemists, working in non-biological areas, wore their lab coats outside the lab, for example, going to the store. They argued that if it was not safe to wear outside the lab it is not safe to wear it inside the lab.

It is often said that breakthroughs occur at the boundaries between disciplines. There were many times when each would call upon the other for expert advice. For example, in chemical safety or handling potentially infectious materials. On one occasion, the laboratory was peripherally involved in determining the cause of a foul odour in a major city building. John Bates, a microbiologist, upon hearing the description of the odour was able to identify the bacterial infection in the wet insulation that was causing the odour. During the anthrax white powder scares both disciplines worked closely to identify the material and maintain chemical and biological safety associated with potential terrorist samples. The amalgamation resulted in a one stop shop for chemical and biological testing.

During this time, a decision was taken to also centralise the receipt of samples, just like in the pathology laboratories. Prior to this, the customers came to the relevant laboratory and delivered the samples, discussed the case with the analyst and learned about other services that were offered. Very close personal relationships developed between regular clients and the chemistry staff. The chemist got a total overview of the background of the sample and was thus able to design the appropriate test. The result of the failure of scientists to have an overview of a case was amply demonstrated in the forensic DNA area, as it did not enable the analyst to detect anomalous results. This led to a commission of inquiry.

At one stage, a single sample entry point was enforced; however, the question was then raised: do you want the bodies for the mortuary to come through the same central point?'. This single sample reception was not implemented.

It was determined that police samples needed different protocols (e.g., continuity of possession to maintain a chain of evidence). While dosimetry samples could come directly to the Health Physics laboratory, dangerous samples associated with a potential terrorism event needed to be treated differently for the safety of sample receival staff.

The old problem raised its head of trying to have one standardised approach for a laboratory that performs a wide range of activities, which vary over time.

Centralised sample receival resulted in the loss of contact between the clients and the technical staff. In 2018, Logan Hospital sent a drinking water sample, related to an off taste, to a private laboratory for a metal screen unaware that the Queensland Health's chemistry laboratories offered a free service. If they had approached the laboratory, they would have been advised to do an organics screen as this was the most common cause of off-taste. When chemists deal directly with clients and know the whole story, better results are obtained. Similarly, there was a move in 2011 not to allow forensic staff to know the background of a case sample on the ground it may bias the results. This move may be appropriate for basic tests such as blood alcohols but not for investigative samples.

MARGARET W
General Manager 2003-2005
Could not be contacted to obtain permission to use name and image.

With the departure of Michael Moore, Margaret W took over as General Manager of QHSS. She had worked at the LMP for many years. Margaret was then in charge of the multidisciplinary public health and forensic laboratories, and she was the first woman appointed to this position. Her prime focus at the time was to reduce the media attention surrounding forensic DNA analysis. The DNA backlog at the time involved tens of thousands of samples. During her time the number of staff employed in this area increased dramatically to over 100 including many temporarily employed to reduce the backlog. New processes were implemented in a production line approach. This was much more efficient compared to the previous slower batch approach where the reporting analyst carried out all the operations. This new approach enabled the work to be streamlined, and the introduction of performance indicators based on production levels. Automated high throughput instruments were purchased, and the Forensic DNA laboratory was significantly renovated to make it suitable for carrying out this new technology.

This improvement in the process was made possible because the Forensic DNA laboratory effectively was only carrying out a single test on samples provided in a fixed form by the one client, the QPS. The police now did the sampling rather than the analysts. This period also coincided with the introduction of the National DNA database with the associated standardisation of test methodology throughout Australia. With the introduction of DNA testing in the late 1980s, the laboratory had proactively accumulated an extensive library of profiles. Unfortunately, as part of the legal framework associated with the standardisation of technology, this local database could no longer be used.

The chemistry areas had, for many years, streamlined processes with instrumental technicians assigned to run batches of samples provided by the analysts. The forensic chemistry laboratories carried out an extensive range of different tests on varying samples for a wide range of analytes. This variety of sample type restricted the introduction of a similar production line approach to the same degree as had occurred in the forensic DNA area. Despite numerous consultancies over 15 years, a better system of handling chemistry drug samples had not been determined. In 2001, the Forensic Toxicology section had about 13 staff. In 2003 Margaret W commenced as General Manager. This number was more than doubled in subsequent years to manage the workload. Similarly, the number of staff in the Clandestine laboratory area increased to overcome accumulated work which stretched back several years. Margaret was instrumental in having the supervisory positions reclassified based on national expertise from PO4 to PO5 salary level. Margaret W left the laboratory in 2005 but continued to serve in the wider organisation.

DOG AND GCMS INTERFACED (DOG-GCMS)

The Veterinary School at the University of Queensland wanted to look at the build-up of a toxic volatile compound during operations on anaesthetised animals.

The dog was anaesthetised in the laboratory, using a recycled anaesthetic gas. Samples of the gas were taken periodically by a line connected directly to a Gas Chromatograph Mass Spectrometer set up to analyse volatile organic compounds.

ROBYN K

Centre Manager Forensic Sciences 2002-2004

Could not be contacted to obtain permission to use name and image.

Robyn K was a Microbiologist who had been involved in the roll out of the AUSLAB information management system. Following the departure of Lenore R, she was placed in charge of the forensic laboratories including forensic chemistry and forensic toxicology. Media criticism continued, mostly focused upon forensic DNA testing and clandestine laboratory backlogs. She commenced the implementation of AUSLAB in the forensic areas.

GREG SHAW
Senior Director 2005-2015

Greg Shaw was appointed as Senior Director of all laboratories at Coopers Plains following the departure of Margaret W. He had previous management experience in the private pathology laboratory sector. The decade before he took on this role, had been marked by extensive, almost weekly media criticism, shortage of resources, several interdepartmental investigations, multiple consultancies, and many managerial and organisational staff changes. Greg Shaw marked a change in management style. Over time, he settled the laboratory down to focus, once again, upon the scientific disciplines of the work and not organisational politics. Upon being appointed, one of his first actions was to restructure the laboratory along discipline lines, so for the first time in 12 years, the chemistry laboratories were back under a single manager. The organisation became known as Queensland Health Forensic and Scientific Services (QHFSS). Interestingly, when the media were asked why they did not mention the lab's name in conjunction with stories, they indicated that it was too long. Perhaps a future name could be shorter. When the Government Analyst or John Tong Centre was the name of the organisation it was frequently mentioned in the media. It was recently noted that when reporting on the dangers of Vapes on the news, the acronym QH FSS (with space) was used rather than the full name but on another TV channel the name was omitted completely, and the reference was only made to "lab results".

During this period, negotiations were finalised with the DPI to construct a new laboratory on the site. Laboratories were also to be provided for the CSIRO Food Division. This construction further extended the range of expertise present on the site. There was hope that there would be considerable collaboration between all the groups on site, and a committee was set up for that purpose. This collaboration did not eventuate to the degree expected. The reason for this lack of collaboration is the quandary between research work and high throughput chemical analysis; when machines are running 24 hours a day, it is difficult to suspend operations while methods are developed, and a limited number of research samples run. In contrast, instruments in research laboratories are often idle and are thus available. The DPI was also separately housed with their own analytical chemists and equipment. This separation further reduced the necessity for collaboration. In the Chemistry laboratory to provide services to the researchers. In hindsight, it may have been better if their analytical chemists were absorbed into the main analytical laboratory.

The replacement and upgrade of instrumentation became possible with the department's Capital Item Equipment Replacement program. Instruments purchased by the laboratory had to be justified in terms of workload or contracts, with a full business case.

During his tenure Greg Shaw developed a submission to increase the staffing of the laboratory; this enabled the appointment of many long-term temporary staff.

The submission also obtained funds for the purchase and update of equipment, thus enabling the laboratory to move into other fields of analysis. Unfortunately, Greg Shaw's work in increasing staffing levels and the capability, of the laboratory was negated in part, by the Cabinet decision to decrease the public service through voluntary, some claim involuntary, redundancies in 2012. There is nothing voluntary about calling somebody into a room and telling them their position is no longer required. The Chemistry sections lost 27 per cent of its staff reducing the number of professional staff from 142 to about 100.

The problem of properly remunerating chemists continued. Past Royal Australian Chemical Institute (RACI) salary surveys indicated that Queensland Government chemists and those in Tasmania competed to be the lowest paid government chemists in Australia. To reduce the loss of talented chemists to other states, Mary Hodge, in collaboration with the Public Sector Union, negotiated and implemented a merit-based progression scheme. This scheme was based upon a submission made by the staff members relating to the level of work they performed. Their immediate managers had to approve the promotion, much like the old Senior Chemist Scheme previously discussed. The merit-based progression system had existed for a short time before a new salary scale was implemented. Only a limited number of staff were successful in obtaining promotion through this process. Greg Shaw was widely respected for his willingness to listen and consult with others. Greg worked at the laboratory until he retired in 2015.

CRAIG R

Acting Executive Director 2018

Could not be contacted to obtain permission to use name and image.

Craig R held an executive-level position within HSQ and was brought into FSS to support the management of the organisation until a permanent incumbent was found.

MORNING AND AFTERNOON TEAS

Environmental health officers delivered samples to the laboratory and a rapport grew between staff. When a centralised sample receipt policy was established, this contact was lost. After a few years, chemistry staff and environmental health officers realised that only the older staff members knew each other, so morning teas were arranged to develop relationships between the new staff. This proved to be a communication success.

At one stage, a problem occurred in the laboratory where the hydrogen flame on a gas chromatograph would not light. At morning tea, the supervisor found out that the wrong gas had been plumbed and there was inert nitrogen in the lines and so the problem was explained.

In the new building there were separate tea rooms on each level. This isolated the staff from communication with other levels. With the expansion of the building program a centralised tearoom was developed incorporating a meals preparation area which facilitated greater staff communication. The forensic labs maintained their own separate tea rooms.

PAUL C
Executive Director 2016-2018

Could not be contacted to obtain permission to use name and image.

Paul Cs was selected in a nationwide search for a new Director of Forensic and Scientific Services. After Greg Shaw retired, the position was examined and reclassified as Executive Director to bring it in line with the organisational structure of Health Support Queensland (HSQ) and to implement executive level contracts of three years.

JOHN DOHERTY
Executive Director 2019-2021

Following Paul Csoban, there was a period when the position of Executive Director was vacant, and Forensic and Scientific Services was overseen by the General Manager of Strategy, Community and Scientific Support, Michel Lok. HSQ undertook another national search and John Doherty was selected. John is a Chemist from the United Kingdom (UK) and had worked in the Australian forensics discipline for several years, including in the Northern Territory and Victoria. He gave evidence under oath at the DNA inquiry that his budget had been cut and he had lost positions despite his objections. This contradicted an executive member of the department who said under oath that the lab could have had more money if they had asked for it. The loss of positions over the past decade has serious implications for the viability of the laboratory.

THE SHROUD OF TURIN

A young woman was murdered on a tourist island. A beach towel was placed over the body. Upon examination a human shape could be seen on the towel. The images were produced by a mixture of paraffin oil, possibly from sun lotion, and soil. The shroud of Turin may have been formed the same way from a mixture of soil and oils.

ARCHAEOLOGY - WRECK OF THE PANDORA

The Queensland Museum requested assistance from the laboratory to identify a brown oily liquid found in the surgeon's chest on the wreck of the HMS Pandora which sank in 1791. It had been sent out to recapture the Bounty mutineers. The liquid proved to be clove oil (Eugenol) used to treat tooth ache. The liquid was still in good condition despite being under the ocean for two hundred years.

GARY GOLDING OAM FRACI

Managing Scientist 2007-2013

With the re-amalgamation of the chemical laboratories, Gary Golding was appointed as Managing Scientist of the Chemical Analysis Laboratories, which constituted the laboratories previously known as the GCL. Gary Golding was a Chemist who commenced working at the laboratory in 1971 as one of Ivo Henderson's bonded scholarship holders. He worked in every section of the laboratory so had an in-depth knowledge of its entire operation. During his seven years in the position, there was a considerable improvement in the backlog of samples in the Forensic Toxicology and Forensic Chemistry areas. Court dates were, at last, being met.

The Health Department had introduced a Capital Equipment Replacement program. During Gary's seven years as Managing Scientist, all the significant items of equipment in the laboratory were replaced. The laboratory also acquired skills in isotope ratio mass spectrometry and was able to recruit Dr Jim Carter, a world authority in the area, from the UK to develop this technology.

The scholarship scheme was reintroduced and attracted young chemists with high academic achievements to the laboratory. Due to the merit-based selection required within the public service, at interview, it was difficult to hire young chemists directly out of universities as they had little experience and were not competitive against more experienced applicants. This process also resulted in an age bulge in the staff with many older chemists and almost no young ones.

To overcome this problem, the scholarship system was reintroduced with the proviso that the person recruited was to be a new graduate in their early 20s. The system operated for two years before the policy was reviewed in 2012 following the mass redundancies. The arguments in favour of this scholarship process do not change over time. Getting the best undergraduates bonded to your organisation before private sector competitors with less formal recruitment process, compared with government, is of long-term benefit to the laboratory. Recruitment policy should include this scholarship system.

As a result of union agitation, the government introduced the Health Practitioners Salary Scale, which extended two levels beyond that which existed in the Professional Officers Scale. Gary Golding was invited by the union to put the case to the Health Department that laboratory staff should be included on this scale. The department was unwilling to cooperate, initially claiming that no other laboratory staff had put in any submission to support the case. Upon returning to the laboratory from the meeting, Gary informed the staff that their future salary classification depended upon them putting in a case. He also informed a member of the Pathology laboratories. The department was inundated by cases supporting the inclusion in the Health Practitioners Scale. As a result of this action, the department agreed promptly, and 50 per cent of the staff received at least one classification level increase in salary. Before this time, the only method of obtaining an increase was to apply for an advertised position at a higher level. These rarely became available. Positions were often downgraded when an incumbent retired.

When terrorism became an increasingly topical issue in the early 2000s, Gary Golding successfully lobbied federal agencies to create a Chemical Warfare Agent Laboratory Network to supply advice and chemical warfare reference materials nationally. The Defence Science Technology Laboratories were successful in obtaining funds to make this possible. The network re-established high-level contact between similar government public health and forensic chemistry laboratories which had been lost when the national directors of similar government chemistry laboratories ceased to meet regularly, partly due to the increased competition for tender work.

In 2012, the Queensland Government decided to reduce the public service by about 10 per cent. Some 14,000 to 20,000 public servants lost their job (the actual number depends upon whether you count full-time equivalents or individual people). As a result of this process, the chemistry laboratories lost 27 per cent of their staff. It was important that the laboratory maintained its capabilities, so the redundancies occurred across all sections. In all, the number of staff dropped from about 142 to about 100 professional staff in the chemistry laboratories. The laboratory could be divided into two groups, staff approaching retirement in the next one to two years and those younger, with less experience, but considerable competence. The future of the laboratory rested with the younger staff. It was considered desirable that redundancies should occur among staff approaching retirement in order that the laboratory maintain its intellectual strength. Should the redundancies have occurred among the younger chemists and then the older chemists retired soon after, there would have been considerable, almost total loss of intellectual capital.

These were dark days for the Chemistry laboratory. In the past, reductions in staff had occurred over time and involved the elimination of vacant positions and staffing freezes. On this occasion, the process was pushed through in a period of a few weeks. To manage this loss of staff, sections were amalgamated and the numbers working in particular fields reduced. The forensic areas, being critical because of continuous media interest, were protected where possible.

Gary Golding retired from the laboratory after 42 years of service, in September 2013, following a restructuring of the laboratory, once again based on forensic and public health lines. The structure was very similar to that previously implemented during the early 2000s. Forensic Toxicology was attached to the Coronial

A truck towing Services area, creating a structure similar to that which existed in Victoria at the time.

30000L ANHYDROUS AMMONIA SPILL

two 33,000 litre tankers of anhydrous liquid ammonia was involved in an accident near the port of Brisbane which resulted in the second tanker spearing into the first rupturing the tank and releasing the liquid ammonia. Laboratory staff attended the scene to provide advice on the clean-up. This was one of many ammonia leaks in the Brisbane area. Most were associated with cold stores. In a confined space ammonia is an explosive gas. It is highly soluble in water and can dissolve in dew on the metal beams in buildings, producing a hazard to those without adequate eye protection as it drips down. The gas concentration can increase rapidly. In one case involving a leak of ammonia, the concentration rose from 10ppm to 240ppm in the length of the analyst's arm.

HIGH BLOOD ALCOHOL

A sample of blood was analysed for alcohol. The result was off scale. Repeat analysis gave a result of 1.6 percent. Investigations found that the body had been embalmed with alcohol injections in Bali. On another occasion a result of 410 mg per 100ml was found. The defendant was to defend the case on the grounds that the analyst made an error because he had medical advice that he would be dead at that level. Prior to the court case, he went on a binge and died from an alcohol overdose.

INDUSTRIAL HEMP

The government decided, following lobbying from industry, to allow the development of an industrial hemp industry. They placed a limit on the amount of tetrahydrocannabinol (THC) in the dried leaf material. The concept was that the level of THC would be so low that there would be no temptation to use it as marijuana. Farmers were to strip the leaves from the stalks and plough the leaves back into the field. One enterprising farmer decided to sell the leafy material and claim when charged that he had a license to grow *it. There were several low THC cultivars available. Some containing almost no THC, others containing levels close to the legislated limit. Unfortunately, one group chose the higher THC variety. The error inherent in the analytical technique was such that this product occasionally failed. The other issue with using the higher THC cultivars was that a smoker could put twice as much product in the pipe to obtain the same effect as the illicit material.*

Some areas have overcome this issue of analytical error by stating in the regulations that the regulated quantity already had a component to allow for analytical error. In other words, if the analyst obtains a result over the regulated amount the client cannot argue that there is a possibility that the real value is below the regulatory limit.

FIRE AT SUGAR REFINERY

In the 1960s the laboratory was asked to assist with the cleanup following a fire at a sugar bulk warehouse. The fire had been put out, but a lot of sugar was in the drains. You would normally expect the sugar to ferment and produce carbon dioxide and alcohol. Somebody noticed that when they threw a

cigarette into the drain there was a small explosion. A bacterium in the drains was decomposing the sugar to produce hydrogen. Sixty years later, a similar fermentation process is being investigated to produce hydrogen.

LEAD AND CADMIUM IN PAINT ON TOYS

Over many years the laboratory routinely carried out testing on the paint on toys and pencils for lead and other heavy metals. Interestingly, on the rare occasions when they heard about it, members of the public regarded the Government's proactive checking of toys in a very positive light. The Health Minister of the time, Mr Austin, took a photo opportunity to announce the banning of failed toys. Chemist Peter White purchased a toy tip truck for his son. The truck was a bright yellow colour. He was suspicious that it might contain cadmium and after testing his suspicions proved correct.

MOLLUSCA ANALYTICUS

Putting the "too hard" sample aside each day, instead of developing a plan to complete it or writing it off, is a problem in laboratories of last resort. Clients expect an *outcome which is not always possible. To encourage one of these actions a section developed a trophy for the chemist with the oldest sample. It bore a cartoon snail with the title "Mollusca Analyticus", "The Happy Gastropod". It sat on the desk of the chemist with the oldest sample. It worked well in the section that developed it but did not translate well to those that did not. In which case it sat on the supervisor's desk to remind them to talk to the person with the oldest sample. This was one strategy that assisted in dramatically reducing the age of the oldest sample.*

CHAPTER 4: GENDER BALANCE

The first, female chemist appointed at the labs was Helen Farrah. She left the laboratory to pursue a career in academia. In the 1960s, women studying chemistry were relatively rare. In the three years (1969-1971) of chemistry classes at Queensland Institute of Technology, now QUT, there were only two women out of approximately 120 students. The first female State Analyst, Cheryl Rynja, was appointed in 1975. Many others followed.

Up until the late 1960s, laboratory staff were predominantly male. During the late 1960s and early 1970s, the number of female chemists increased. The ratio gradually changed over the next 40 years until there was approximately an equal number of males and females employed at the laboratory in professional roles. There were several impediments to women working in chemistry generally. Initially, they only received 85 per cent of the male wage for the same work, though this situation was changed in the mid-1970s. Women were not allowed to continue working after they were married or even allowed to wear long pants to work.

Standards of behaviour have changed. At that time (early 1970s), the double entendre was accepted and widely practised. With a predominantly young male cohort of staff, sexual harassment of female staff was witnessed, usually involving the same small group of individuals. This misconduct was not restricted to the younger staff members. Some former female employees still talk in damning terms of the conduct of one male senior staff member. Even during lectures at university one female staff member, Lenore Hadley, was questioned as to why she was studying chemistry; she responded in her own way by 'being top student in that class.' She went on to become the first person to graduate with an MSc(hon) from Griffith University in Forensic Science.

Although increasing in numbers, professional women chemists and technicians were in the minority at GCL in 1982. The few professional women inspired younger women. In 1982 pay and conditions for graduate women and men were generally equal at the laboratory, however, the executive and leadership level staff were essentially entirely male until much later.

Maternity leave without pay was available in 1982, however, women could only take off a maximum of one year of unpaid maternity leave and then had to resume work full-time. Part-time work was not available unless a supervisor drew up a special employment contract for the employee; there was only one example at the laboratory in 1992. This situation resulted in many highly skilled women leaving the workforce to care for their children. The few women returning to full-time work often used long service and recreation leave to cover periods of parental duties. There was initially no paid paternity leave until 2011 when paid parental leave was introduced.

Formal childcare placements were difficult to obtain as centres were much scarcer than they are in 2022. In 1993, a part-time work arrangement with fixed hours and limited tenure (until their child was two years old) was made available to women returning after their maternity leave.

Eventually, access to part-time work became as flexible as full-time work and was available on a permanent basis. The increasing flexibility of part-time work for parents has resulted in many more women remaining in the workforce and retaining their skills.

CAFFEINE IN RACEHORSES

A scandal erupted in the racing industry in the mid-1980s concerning the detection of caffeine in the urine of racehorses. Owners strenuously denied being involved in doping their horses. The GCL was called in as an independent third party to investigate the issue. Following a considerable effort by Chemists Neville Bailey and Geoff Eaglesham, they found that the caffeine was a contaminant from the test strips used to test the urine pH prior to the extraction and analysis of the horse urine samples for drugs. Such was the impact of their findings clearing many a horse trainer, that the then Minister for Racing, Russ Hinze was very grateful and took them out to lunch. The Courier Mail labelled them back-room boffins.

DEALING WITH REDUNDANCIES

The saddest part of any manager's job is when they are directed to cut staff, 99% of whom are good workers. The consultants asked what would happen to the work if a particular person was made redundant. The answer is that we would have to stop doing that work knowing the client would complain to the Minister. A better answer is, we would have to outsource the work, i.e., we will have to put the work out to tender. It was noted during the 2012 redundancies process that some positions were to be made redundant pending outsourcing of the work. The tender process did not attract any interest, and the positions remained with the laboratory. Similar things happened elsewhere in the government. As the laboratory of last resort, there was nobody else to do the work so outsourcing would not succeed.

SUICIDES IN MOTOR VEHICLES

It was common to receive samples of blood from people who had committed suicide in cars using the car exhaust which contained carbon monoxide. The blood contained carboxy haemoglobin. With the introduction of catalytic converters less carbon monoxide was produced by motor vehicles and tests showed that the exhaust gradually displaced the oxygen in the car, so the person effectively suffocated. Other suicides involved the generation of hydrogen sulphide which is very toxic but fortunately has a low odour threshold

WHITE PHOSPHORUS MATCHES

The ban on white phosphorus matches was removed from a regulation on the grounds that there are no white phosphorus matches in the marketplace. It seemed to have escaped the regulator's attention that this may have been due to the ban. These matches had a habit of catching fire in a person's pocket. This ban was an international convention ironically called the Berne Convention (1906).

CHAPTER 5: INPUT INTO LEGISLATION

The laboratory's role does not just include analysis but has a significant advisory component. It continues to have input into reviews of Acts of Parliament.

The laboratory has provided feedback and input into several Acts of Parliament (State and Commonwealth). Input included:

Crime and Corruption Act 2001

Drugs Misuse Act 1986

Criminal Code Act 1899

Criminal Justice Act 1989

Dangerous Goods Safety Management Act 2001 (repealed 2012)

Drugs Misuse Regulation 1987 (Qld),

Industrial Cannabis production

Evidence Act 1977

Fire and Emergency Services Act 1990

Health Act 1937

Justices Act 1886

National Measurement Act 1960 (Cth)

Pest Management Act 2001

Public Health Act 2005

Food Act 2006

Water Fluoridation Act 2008

Tobacco and Other Smoking Products Act 1998

Transport Operations (Road Use Management) Act 1995

Coroners Act 2003

Public Safety Preservation Act 1986

Quarantine Act 1908 (Cth)

Biosecurity Act 2015 (Cth)

Agricultural Standards Act 1994

Workplace Health and Safety Act 1995

Chemical Usage (Agricultural and Veterinary) Control Act 1988

ADVICE ON LEGISLATION

Geoff Rynja recalls that the lab had input into the drafting of the Drugs Misuse Act 1986. He attended the meetings with the Parliamentary draftsman. The Queensland Government proposed that if a person was found in possession of 0.5g of powder and the analyst found the powder contained any heroin, then the person was in possession of 0.5 grams of heroin. At least, he was able to convince them to go with the actual quantity of drug.

CHAPTER 6: LABORATORY BUILDINGS

Early history is taken directly from "A History of Health and Medicine in Queensland," 1824 to 1960 by Ross Patrick (1987). It includes additional editorial comments.

The laboratory had several locations and changes in the portfolio under which it was administered. When Robert Mar was about to take up the position in 1882 it was announced that he would probably be accommodated in the old grammar school building in Roma Street. Within the space of a few years, post office directories gave other addresses including the Post Office Building in Queen Street, School of Arts in Ann Street and then William Street. The William Street site was in a building on the Victoria Bridge side of the old public library building which housed the State Children's Department. In July 1905, the laboratory was transferred to a new location in the former Executive Building in George Street next to Queen's Gardens. A further shift to an extension of the Department of Agriculture and Stock building took place in 1935. The laboratory stayed in this location until the move to purpose-built laboratories in Coopers Plains in 1989.

The position of government analyst was attached to the Treasury Department when Robert Mar was first appointed in 1891. The laboratory was transferred to the Mines Department and remained under its jurisdiction until 1896. Then followed a short-term with the Home Secretary Department as a branch of the Health Department until September 1907 when the Treasury Department again resumed control. Finally, the laboratory was one of the services included in the transfers to the Department of Health and Home Affairs as part of the reorganisation in 1936. It has remained with Queensland Health since then.

In the 1960s the central laboratories were situated in half of the subbasement of the building in William Street next to Queens Wharf Road. A small office area and administrative section occupied part of the basement floor.

The laboratory also occupied a building on the corner of Alice and William Streets opposite Parliament House. This building was used for water analysis and some analysis of commercial products. Half of this building was demolished to make way for the ramp to the new Southeast Freeway in 1970. During the construction of the freeway, it was almost impossible to use the balances as there was significant vibration due to the pile driving necessary to support the freeway. The laboratory was generally known as 'Siberia' as the work was routine, and it was perceived as a punishment to be sent there to work. Later, the remainder of the building was demolished, and it was used as a government car park from 1982–2013. It is now the site of number one William Street, a large high-rise building housing public servant. The main William Street building had no sign or street numbers. This was done on purpose to improve security.

The Waters laboratory moved from Alice Street into the main laboratory complex when the DPI vacated its laboratory space for their new laboratory at Yeerongpilly. The laboratory occupied the entire sub-basement of the building and part of the basement floor. Later, a forensic toxicological laboratory was built on the basement floor near the front of the building. By modern standards, workplace, health, and safety conditions were not optimal: fire assays for gold and silver analysis were carried out in part of the building. Solvent fume extraction systems were not adequate, fume cupboards were constructed of wood and glass, and signage was kept to a minimum to make it difficult for thieves. There were no on-site security guards and there were several break-ins in the building. An arsonist also tried to set fire to the fire assay room, furnace room, library and one of the toxicology laboratories.

In the mid-1970s the building was found to be infested with the West Indian dry wood termite. This infestation necessitated fumigation of the entire building along with many others in the precinct around George Street. It was necessary to move the laboratory equipment and staff to other laboratories in the Brisbane area so work could continue during the six weeks of the fumigation process. The laboratory played a role in monitoring levels of methyl bromide in the buildings and the blood of the workers

Initiation – during the fumigation of this and other buildings in the central business district.

THE COOPERS PLAINS LABORATORY AND STRUCTURE

PLANNING

The building in William Street was in a deteriorating condition which limited the growth of the laboratory.

Air conditioning was provided only for critical instruments and windows had to remain open for ventilation. Situated as it was, next to the freeway, contamination issues became critical as airborne particles entered through open windows. This was particularly problematic in the trace metals area. Previous attempts to eradicate termites had resulted in a background level of heptachlor in the atmosphere of the laboratory. This limited the analysis of trace organochlorine pesticides. Overall, the laboratory building and the infrastructure did not conform to the current standards for laboratories. Rats were also a problem. The drain of the building led to the river and rats were known to frequent the facility during the night. On one occasion they ate the entire contents of a loaf of bread which had been sent for analysis. On another occasion, the rats knocked over a container of chromic acid (concentrated sulphuric acid and potassium dichromate). Footprints of the rat were seen leading away from the spill. In the Food laboratory, rats gnawed through a skirting board and created a classical hemispherical rat hole in the wall. Appropriate eradication was carried out which resulted in the capture of nine large rats in one weekend. For a period following the fumigation, the smell of decaying rats permeated the building.

The laboratory was in a flood zone. During the 1974 Brisbane flood, water rose up to the back steps of the laboratory and flooded some of the rooms under the subbasement and outbuildings. A rise of another 20cm would have inundated the building and destroyed major items of equipment. Staff who were able to access the city came during the weekend to put equipment and reagents up on the bench tops. Following the Australia Day Public Holiday, staff commenced the clean-up of the grounds and outbuildings. A digitised colour movie film taken by the author, showing the flood around the laboratory, is available from the State Library site.

In the 1970s, one wing of the sub-basement was renovated to house newly acquired instruments. These included atomic absorption spectrometers, an x-ray fluorescent spectrometer, several gas chromatographs, and an emission spectrograph.

Trevor Beckmann became the Director of the laboratory in 1980 and he initiated action to have a new laboratory building placed on the Government's Capital Works program. Over several years, the laboratory moved up the priority list and in 1983, planning began in earnest. Don Lecky, assisted by Gary Golding, was assigned the task of coordinating the planning for the new laboratory. Funding of $25 million was allocated for the construction and fitting out of the new laboratory. This amount was not based upon any cost assessment but on it being perceived as being equivalent to a small hospital.

PHILOSOPHY OF CONSTRUCTION

It was recognised that the work methods of the Government Analyst had changed over time and would continue to change, and the laboratory had to be built with a high degree of flexibility to accommodate these changes in practise. It was decided that to reduce costs, the laboratory should be open plan, rather than a series of small laboratory rooms. This had the added advantage of creating more space. A large room needs one corridor, two small rooms need two, often on the other side of an adjoining wall. Unfortunately, this philosophy was lost in subsequent years as small rooms proliferated.

Don Lecky visited laboratories interstate and overseas to gather ideas on infrastructure. Initial planning based on a 50-year growth projection resulted in a building that far exceeded the budgetary allocation; considerable cutting back was therefore necessary. The simultaneous design and construction approach meant that decisions were required continuously and could not be changed once construction commenced. During construction it was found that desk mounted power outlets did not conform to the standard. A committee did a risk assessment based upon past experience that there had not been any fires caused by desk mounted outlets. Construction continued. There was extensive consultation between laboratory staff, laboratory management and the architects in the design of the building.

The new laboratory was planned with flexibility in mind to accommodate the growth and changing needs for analytical chemistry work for the next 50 years.

 As a result, based on the previous 20 years of growth, the laboratory was much larger than the one in William Street and not fully utilised by the GCL. The final occupancy coincided with a recession and thus, an absence of funds for additional building works including parking facilities for the staff. As a result, the excess size of the laboratory was later able to also house the Government's Health Physics laboratories, the Laboratory of Microbiology and Pathology, the Police DNA unit, the AUSLAB team and a University of Queensland research laboratory within the original building. Only 15 to 20 years after construction some chemistry sections were once again seriously overcrowded. Sections that initially accommodated 10 to 15 people now accommodated over double that number.

 By 2014, the chemistry laboratories had been restricted to one block (six laboratory wings) of the building. The original four chemistry laboratories in block three (four wings) had been taken over by others and the chemistry groups restricted to six wings in block one. Over time, more and more small laboratory rooms were built, resulting in inefficient use of space. The need for administration areas within the laboratory sections became imperative with the introduction of more computers.

Analysts now typed their own reports, and the centralised typing pool disappeared. A considerable amount of time in each analysis also involved paperwork, quality checks and peer review. Office areas were expanded at the expense of laboratory space.

A vital feature of the building's design was service corridors between laboratories and along the outside walls. These service access points necessitated peninsular benches but enabled any services to be delivered to any bench in the laboratory with minimal disruption.

As previously mentioned, the William Street laboratory had air conditioning only in the instrument facilities. The new laboratory was to be air-conditioned throughout. However, a large number of fume cupboards were necessary to ensure the extraction of hazardous materials.

The combination of air extraction and air-conditioning resulted in the decision to implement one pass air-conditioning, rather than the traditional eight to ten changes per hour. This high air replacement rate significantly increased the costs of running the laboratory in terms of electricity charges. The combination of one pass air-conditioning and fume cupboards resulted in a rather noisy background sound in some areas of the laboratory.

The flexible nature of the design and the open plan meant that when parts of the LMP business were incorporated into the building, changes could be easily made.

In the mid-1990s renovations were underway in some of the laboratories. To make them suitable for microbiological testing, bench modules needed to be removed and replaced with a solid form bench. Benches containing gaps created potential for contamination within microbiology laboratories. Although a proposal to reduce the size of the chemical laboratory was put forward, it was fortunately not actioned and in 1998 virology laboratories moved into a new purpose-built wing of the building. The funding of a new major laboratory building recognised the importance of chemical analysis to the government. Being situated in the city, it was a convenient location for staff from all over Brisbane. It was rare during that period for anyone to resign from the laboratory. This staff stability was partly due to the absence of analytical chemistry positions in other laboratories in Brisbane and staff satisfaction.

EXPLOSION AT POST OFFICE

An explosion occurred at the GPO in Brisbane in the mail sorting area. The source of the explosion was identified as a package of starter caps for use in athletics starter pistols. Perhaps the package had been thrown.

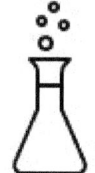
LOCAL GRAFFITI

There was a considerable difference in education requirements between some of the newly graduated degree chemists and older managers who had sugar diplomas. In the William Street male toilet one wag placed a sign above the toilet paper stating: "Get your sugar diploma here." On another occasion an extensive scrabble game was also played on the same wall with the names of several managers filled out with many derogatory words. One manager was so impressed he photographed the "artwork."

BLOOD ALCOHOLS AND ANALYTICAL ERROR

The Government sets a limit of say 50mg /100ml for blood alcohol. What does the analyst report if the triplicate results are 50, 49.5 and 50.5. If they analyse the sample 30 times, they can calculate the standard deviation or the expected spread of results. They overcome this problem by subtracting an amount that gives them 99% confidence that the result is above the reported number. It is better to use the term analytical variation than error, lawyers love errors.

DON'T ASSUME

The drawings of the under-bench units were presented and approved by staff. Unfortunately, we did not ask to see a drawing of what was inside the cupboards. The answer... no shelves, they were just boxes. The lab over time retrofitted shelves to double the capacity of the cupboards. There were not enough funds left to landscape the grounds. The staff did landscape at the entry and the end of block 3. Unfortunately, further building works demolished their good work. The car park was sealed as part of future building work. Initially there was no turn in lane off Kessels Road and staff had to be careful there were no trucks bearing down on them when they slowed to turn into the entrance.

GLOVES WITH UNPLEASANT ODOUR

The University of Queensland submitted a pair of gloves that had a foul-smelling chemical on them. The compound identified occurred naturally in a particular herb an extract of which was being transported in the same shipping container.

THREE KILLED IN QUEEN STREET

At a building site in the central business district, contractors were placing the last facing stone on a high-rise building when the nylon lifting sling, they were using broke. These slings had a loop on either end and were used to lasso objects to be lifted by cranes. The slab fell onto the sidewalk killing three pedestrians. The sling was sent to the laboratory to ascertain the

cause of the failure. It was decided to test the other unbroken end of the sling at QUT on a tensile strength machine. The specification required the sling to safely support twelve tonnes and be tested to twenty tonnes. The other end broke at approximately one and a half tonnes. This demonstrated a fundamental flaw in the product and resulted in a change in the Australian Standard and the use of chains on cranes. It became known during the investigation that these slings regularly broke. As a wise person once said, "there was never an accident that did not announce it's coming." Near misses are especially important. Concerningly, similar slings are still used overseas, in one case to lower David Attenborough's submarine into the ocean.

THE ARCHITECT EXPRESSES REGRET

The architect who designed the laboratory complained that there was only one curve in the entire building. It was in the director's office. He complained that the only place he could exercise artistic talent was in the main foyer. We teased him by saying he designed a void.

The new building provided an opportunity to break up some of the larger sections and amalgamate other sections into a new organisational structure.

SPACE PLANNING

The building consisted initially of two laboratory wings, an administrative wing, a services area, and a staff facility central block. Other wings have been added over time including a virology laboratory, a separate building for forensic biology, pathology and the DPI and CSIRO laboratories.

CHANGING STAFF NUMBERS

Staffing has remained remarkably stable throughout the laboratory's history with very few staff resigning to take up positions elsewhere. Many staff worked their entire lives in the laboratory. However, the move to the new site led to an increase in turnover. People residing on the north- side of Brisbane had to decide if they would stay at the laboratory or seek employment elsewhere. Some moved to the south-side to be closer to the laboratory.

The GCL had a staff of approximately 110 professional and administrative staff when it moved to the facility at Coopers Plains in May 1989. Chemistry staff grew to approximately 142 professional and technical staff in 2012. This number decreased to approximately 100 professional staff after redundancies in 2012.

INCONVENIENCE

At the time, the relocation of the laboratory to the suburb of Coopers Plains resulted in some inconvenience to the staff. Public transport was not optimal, particularly for staff living on the north side. There was no nearby shopping area for banking and the purchase of food. However, car parking was provided on site free of charge. A cafeteria facility was developed.

ONE-PASS AIR-CONDITIONING

One-pass air-conditioning became an issue when the whole building turned on at once at 6am. The electricity was billed based upon peak load. Dan McKeown successfully arranged for a staged turn-on to reduce the peak load. Later as a part of wage bargaining, we were told to make electricity savings. We were not allowed to include the large amount saved by the staged turn on, so we had to resort to turning off lights. Rooms were fitted with movement sensors that turned off the lights when there was no motion in the room. A bit of a problem arose when the lights went off during a meeting. People got used to waving their arms wildly when the lights went out.

GAS LINE CONTAMINATION

When the DPI laboratories were being built the instrument gas lines were contaminated by the oil used to draw the copper pipe. We were determined to avoid this problem, so the gas lines were all cleaned prior to installation. Unfortunately, the plumbers then used a Teflon/ oil-based product to seal the connections which resulted in gas contamination. The gas lines had to be removed and washed in a solvent. Unfortunately, the solvent was contaminated resulting again in contaminated gas supply pipes. Finally, the pipes were washed then heated on a rack of gas burners. Dan McKeown solved the problem. We fortunately avoided other contamination problems including the use of imported plywood treated with pesticides for the construction of laboratory cupboard.

ORGANISATIONAL STRUCTURE CONSIDERATIONS

When considering organisational change many factors must be considered.

- Structure follows from strategy. What is the strategy – does it fit all sections.
- Is there something wrong with the current structure - define the problem.
- Ensure there are sufficient operational, tactical, and strategic levels of management to deal with these issues
- Some sample types require specific facilities, handling, and infrastructure (e.g., biological)
- The levels being determined are critical (e.g., you cannot analyse pesticide formulations in a trace pesticide laboratory)
- Specific instruments can analyse many different types of samples (e.g., elemental analysers are not particularly concerned with the source of the digested sample)
- The layout of the building influences workflow, so it is more efficient to carry out certain jobs in a particular location
- Having a client focus is essential to build up a relationship with the client
- There is no point setting up a laboratory to perform a specific task if you do not have the individuals with the necessary discipline skills.

- It is very difficult to carry out long term research in a short-term factory environment.
- Management consultants often focus upon one aspect, usually a client focus, to justify restructuring the business. Over the years, the names of the sections have changed but they still perform the same tasks.
- Every consultant we encountered demonstrated their worth by restructuring whether it was justified or not.

VARIATIONS IN ORGANISATIONAL STRUCTURE OVER TIME

STRUCTURE- 1975

At this time, the GCL consisted of a waters section, an ores section (mineral analysis and industrial hygiene), a foods section, a forensic toxicology and illicit drug section, a customs laboratory (customs and drugs and poisons analysis), a commerce laboratory (government contracts and analysis of commercial products) and an instrumental section. The management structure consisted of a director, a deputy director, four (later reduced to three) chief chemists and five supervising chemists.

STRUCTURE—1989

With the move to the new laboratory at Coopers Plains there was an opportunity to develop additional chemistry sections. The building had ten chemistry laboratory wings. A structure was developed that occupied each of these wings. The Food laboratory continued much as it was. The Waters section was divided into two parts: one performing organic analysis and the other performing inorganic analysis. The Mining laboratory split off a section known as Occupational Hygiene. The Forensic areas were divided into Forensic Toxicology and Forensic Illicit Drug Analysis (also including physical evidence). There was a food commodity and Food Quality section including a Pesticide section, a Commercial Products section (government contracts and drugs and poisons analysis for Queensland Health) and a specialised instrument area.

In the early 1990s, a decision was taken to move the occupational hygiene chemists to another branch of government. The laboratory lost most of its capability in this area although some work was still carried out within the Manufactured Products group.

With the introduction of the NRCET/ENTOX group into the laboratory, areas previously occupied by Occupational Hygiene and the GCMS/HPLC service section were allocated to this new group. Over time ENTOX also took over the service areas, and the Instrumental section was reallocated to the Organics area and part of the Manufactured Products section. The management structure consisted of a director and deputy director, three assistant directors (chief chemists) and eight supervising scientists.

MEAT PIES

Meat pies were routinely tested for conformity with regulations, sometimes with unusual results. In one case, a cube of hide with hair still intact was discovered. If a piece of vegetable was present, e.g., corn, it became a meat and vegetable pie and was not covered by the meat pie regulations

FLEXITIME

This was introduced in 1977 and quickly became ideal for laboratory workers. Laboratory work cannot be put down like a pen when the clock struck 5:00 pm. Often staff would not commence any new work after 4:00 pm if they could not get it to a stable state before 5:00 pm. Following flexitime, the spread of hours that staff could work was extended from 9:00 am to 5:00 pm to 6:00 am to 6:00 pm. This greatly improved instrument utilisation and productivity. Staff no longer had to ask permission to leave early to attend to personal matters.

DOG BAITS

Baits were laced with broken glass, snail baits, strychnine, cyanide, rat poison, fishhooks and 1080 poison. It is important to only expend resources if there is a likely outcome. An ongoing problem with the analysis of suspected dog baits is that the analysis is very expensive and the possibility of a successful prosecution very low. Even when the person was seen throwing the bait to the dog, successful prosecution was difficult. They would claim that 'they thought the nice dog may like it.' Cost is an issue in cases where there is no visible sign of poison. Visible amounts can be quickly identified by infrared spectroscopy. It was finally agreed at senior management level with the Police and the Health Departments that dog baits would not be analysed. This decision did not hold as at operational level, police felt that the samples should be investigated, as they posed a danger to children.

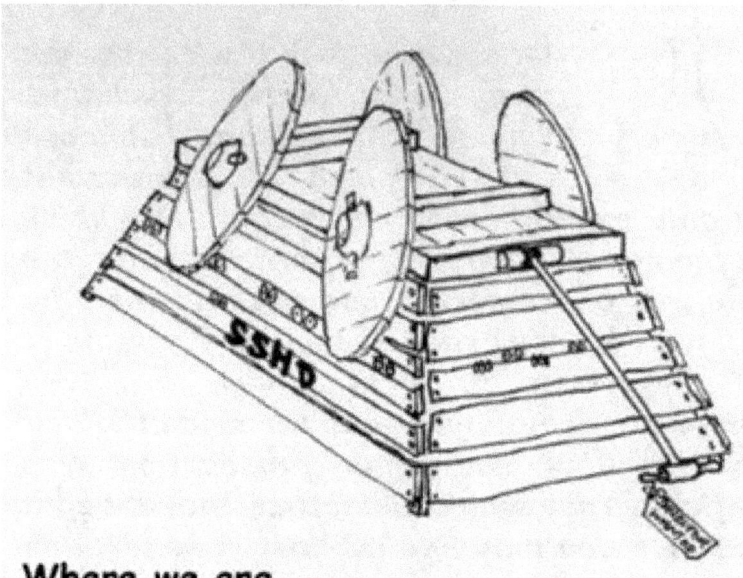

Where we are

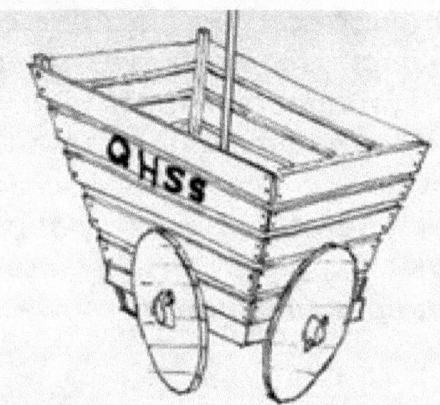

Structure designed by consultant

One manager's view

WHERE WE SHOULD BE

STRUCTURE—1996

(*Drawings courtesy of George Lee* loosely based upon the tumbrel used to take prisoners to their execution in France.)

This restructuring was driven by the need to incorporate the microbiologists from the LMP into the existing building. The presence of non-chemists within the structure led to the development of a Public Health branch and the Forensic branch.

During this restructuring, the Environmental Water section was combined into the Pesticide Residue section to form Organics. The inorganic waters work was reassigned to the Mining laboratory renamed Inorganics. The structure focused on the efficiency of processes, not outcomes or value-added advisory services. However, for a period the Food Trace Metal section remained in the Food laboratory.

Around this time, health physicists who carried out radiation testing were moved from Radiation Health and incorporated into the Manufactured Products group which had been renamed Investigative Chemistry to show it had a broader role. The regulators at Radiation Health perceived conflict between regulating and generating evidence for prosecution within the Regulatory branch. When providing evidence to the courts, the independence of the laboratory was seen as positive, in much the same way as the forensic laboratories are independent of the police. The structure now consisted of a Director, A Deputy Director, Two Branch Managers and Six Supervising Scientists

STRUCTURE—2007

This structure was preceded by a ten-year period when managers were not required to be subject matter experts in the area they were managing. This approach caused some problems in the forensic area where it was a NATA requirement that the manager had a certain number of years' experience in the forensic field. The position of scientific manager was introduced to assist with this requirement.

With the arrival of a new senior director in 2005 to run the vastly expanded laboratory, a decision was taken to structure the laboratory along scientific discipline grounds and to appoint managing scientists, who would be subject experts, to run these different disciplines. The new chemistry structure included a Forensic Toxicology section, Forensic Chemistry (illicit drug and physical evidence), an Organic section, an Inorganic section, Food section, and Investigative Chemistry section. The expanded laboratory's organisational structure consisted of a senior director, a deputy director, one managing scientist of chemical analysis and six chief chemists.

STRUCTURE - 2014

In 2014, following the 'voluntary' redundancy of a large percentage of the chemistry workforce, the laboratory was once again restructured under the disciplines of Public and Environmental Health, Coronial Services and Police Services. This structure was influenced by the client groups the businesses serviced. Public and Environmental Health contained Organic and Inorganic Chemistry, Public Health Microbiology, Public Health Virology, and Radiation and Nuclear Sciences. The Police Services group included the laboratories of Forensic DNA and Forensic Chemistry which included the physical evidence and clandestine laboratory groups. Coronial Services looked after the mortuary, counselling, and Forensic Toxicology laboratory. Managers were once again not discipline experts.

Over time, the administrative area became an entity in itself because of a substantial increase in the size of the establishment and the need to manage laboratory buildings.

The number of senior middle managers in the chemistry laboratories was reduced from four (director, deputy director, three chief chemists) between 1970–1995 to two (senior director and the managing scientist chemical analysis in 2007). In this case, the senior director oversaw the entire business, not just the chemistry laboratories. Effectively the number of senior chemistry positions was reduced to one. Because of the workload associated with a staff of over 142 people, it was difficult for the managing scientist to maintain close contact with staff.

Similarly, the senior director was seldom seen in the laboratory. The loss of middle management positions inhibited ambitions for career progression and opportunities for succession training of senior managers. The growth of the administrative area made the management of such a large group possible by removing some of the day-to-day tasks from the managing scientist.

The merger with the LMP had positive and negative consequences. This merger necessitated a name change with the disappearance of the GCL as an independent entity. The cultures of the two organisations were somewhat different primarily driven by the differences between the disciplines of chemistry, pathology, and microbiology. There was some limited conflict between practises.

On the positive side, it enabled the concept of a one-stop-shop to be available for clients seeking chemical, radiation and microbiological analysis.

The merger into a larger organisation which included the Pathology Queensland laboratories led to new possibilities. It was possibly the first time the laboratory had a direct chain of functional management communication to the highest levels of the department. A Senior Manager was brought in to run this larger organisation, as the Chief Executive Officer had previously privatised a large Commonwealth hospital. This history led to suspicions that a similar fate was planned for the laboratory, but this did not prove to be the case.

The merger of the laboratory with Pathology into Queensland Health Pathology and Scientific Services gave access to funding that enabled the re-equipment of the laboratory and other organisational efficiencies associated with a larger organisation.

The environment changed from one where decisions were often made informally, to an environment where decisions had to be accounted for in great detail. Previously, a few lines of justification were all that was required for major purchases. This situation changed to require full business cases and documented contractual processes often run by a central government agency.

The chemistry laboratories, used to taking charge of their own destiny, were being directed by the larger organisation to fit within a mold. Initiatives developed by the chemistry laboratory were adopted by the larger organisation, usually delaying a local implementation for years so that they could be adjusted for the organisation's needs. These adjustments were often not optimal for the chemistry laboratory.

There are considerable parallels with the Lockheed Skunk Works concept. Under this concept, engineers from the production line who are used to solving problems within a few minutes were placed in research laboratories with people who are used to taking years to solve the problem. The result was a dramatic improvement in research outcomes. Unfortunately, the chemist's ability and problem-solving skills were not recognised by the larger organisation and the opportunities to increase efficiency and problem solving were limited. Organisational politics drove change.

As previously mentioned, the chemistry laboratories had always been not strictly aligned with Queensland Health core business and the staff related more to the GCL than to the Department of Health. In a few short years, the GCL became fully integrated into the wider organisation. The number of ministerial information requests changed from a few per year in the 1980s to almost a constant stream in the 2000s. Staff were shielded from most of these issues although there were many vigorous corridor discussions referring to the bureaucratic system. For many staff, the changes in names and structure had very little impact on their day-to-day work.

CONSULTANTS

Several types of consultants were employed in the 15 years from 1995. The first ones did a course in business planning and oversaw the restructuring of the chemistry laboratory. When consultants were helpful and willing to discuss change and make suggestions, they were well accepted. Once they crossed a line and became dictators of change, they

experienced resistance. Good people skills are essential in consultants. In one case the staff tearoom was cut in half to make room for the consultant's office, bad move number one. Then the consultant came out in the middle of morning tea and told the staff to keep the noise level down as he and his team were trying to work, guaranteeing resistance. Another consultant was obsessed with generating numbers on every aspect of the work and having them reported to him daily. When he asked why the number of autopsies had gone down the answer was, fewer people died!

Another consultant, during a large meeting of staff, told the audience how bad the previous managers were. Most of these previous managers were in the audience. After the meeting they confronted him and asked him to clarify his remarks. He could not. Apparently, there had been some undermining of previous managers by an incumbent. As part of an MBA the author read about a new change manager saying the previous management did a good job as the organisation was still here and thriving. However, things will be changing. He immediately had the previous managers onside. At one point leading up to the new extensions, staff were required by the architect to flow chart all their operations. A *huge amount of time was spent doing this and no outcome was ever visible to the staff. A few years later they were required to do it again. Meanwhile the only performance indicator that concerned staff, the backlog of work, grew. A consultancy related to a new LIMS system was well **managed and staff consulted. Unfortunately, this did not result in a new system to replace AUSLAB.***

CHAPTER 7: INTRODUCTION OF USER PAYS 1983

Beware of the client who does not ask the price. He has no intention of paying.

Gary Golding 1995

Up until the early 1980s, the GCL charged for only a few laboratory services, mainly rare requests from the private sector. In 1983, the Treasury decided to move the laboratory down the user- pays path. Traditional block funded clients continued to be funded from the Treasury. Clients from other departments and clients requesting new work would be charged. During the years from 2007 to 2012 attempts were made to negotiate a service level agreement with the Queensland Police Service (QPS). This stipulated that if the value of analysis exceeded a certain budgeted amount the QPS would pay for additional tests. It became apparent that the QPS were unwilling to change the existing service arrangements.

A similar situation occurred in an overseas lab. This lab also provided a salutary example of the hazards of a user-pays approach in the forensic area (see Appendix 22). A user-pays system does not allow managers to plan for significant equipment purchases, permanently hire staff and build their expertise. It is patently clear that an expert witness must have expertise. Massive backlogs developed because of these constraints, and after major judicial and media pressure they developed a system like that which exists at any laboratory that receives government funding. The Queensland commission of inquiry into forensic DNA emphasised the importance of reliable results over budgetary issues.

Client-specified tests works well when testing a suite of similar samples (e.g., blood from a human) but do not work well when there are variables in the sample type. The clients decide on the tests required, rather than the analyst. This system works well in areas like Forensic DNA and medical Pathology labs, which have several standardised tests on a consistent matrix. In the chemistry area, every sample is different and may require a different approach. Interfering compounds are common. For example, as mentioned previously, one herbal medicine was analysed by ICP for a suite of heavy metals. Uranium was indicated. Radiation tests did not detect it. If the test had been automated, without overview, we would continue to believe that uranium was present, when high levels of iron had interfered and produced a false uranium signal. Such errors can have profound consequences.

The chemist is an investigator, who must consider a wide range of issues based on evidence and vary the process if, or as, required. This process is difficult with fixed price analysis. The client needs an accurate quote for the work performed. It is often necessary to explain to the client that chemical analysis is an investigation not dissimilar to going to the doctor for a diagnosis. Some medical tests will produce negative results, and more tests are required; you still must pay for negative results.

If a client asks for a specific test with a suite of analytes and the analyst finds additional components present (peaks in the chromatogram), under a user-pay system the chemist cannot investigate further, and it may not suit the objectives of the client to obtain the additional information. For example, in one case the client asked for a specific test which involved the detection of benzene in ethanol. This was a standard British Pharmacopoeia test, and the analyst noticed a rise in the baseline of the spectrum at the end of the specified spectral range required by the method. Further analysis determined the problem was contamination of the sample by phthalates from the liquid transfer tubing used by the client. Failure to look further into the rising spectral baseline could have resulted in the rejection of the product.

There are many areas where user-pays does work, especially with standard suites of analytes (e.g., a pathology laboratory). However, in a 'laboratory of last resort' where a problem must be solved, limiting the budget for the test could have deleterious outcomes.

It was also challenging to operate commercial or contracted analyses within the Department of Health as only the Director General could contract the department to do work. Getting a suitable contract to be signed by the Director-General was a very lengthy process and clients were unwilling to wait. For example, on one occasion an entrepreneurial group approached the laboratory to perform a clinical trial on a new drug they had developed, but due to bureaucratic constraints, the two-week timeframe for the signing of a contract could not be met. Business processes that were historically managed internally were now a part of the Department of Health. On one occasion, the laboratory generated their own business cards and were admonished personally by the Director General at the time.

On another occasion, a client sent a sample consisting of royal jelly in capsules for analysis. The claimed weight of royal jelly on the label was several times the actual weight of the capsules. A certificate was written, and the disgruntled client complained and brought this to the attention of the then Director-General. The laboratory was required to explain the reason for taking on commercial work even though it had been doing it for many years with the approval of the department and as a requirement of the Treasury. An attempt was made in the mid-1990s to set up an internet presence, so the laboratory could take advantage of a system that would allow clients to access their results online. At that stage, the department was very risk-averse and did not allow the internet on the network. These were major hurdles in the way of an efficient, commercially viable organisation.

GROWTH OF USER PAY

User-pay or fee-for-service income grew consistently and in 2012, approximately one third of the chemistry budget was derived from fee-for- service income. As the user-pays income grew there were fears that the laboratory would become a target for commercialisation. In the early 1990s the government decided to commercialise any unit with a budget of more than $14 million. The amalgamation with the LMP put the laboratory in this range. A series of hung parliaments saved the newly amalgamated laboratory from that fate. The growth in user-pays income occurred when major government initiatives were undertaken such as the recycled and desalinated water projects. The drought of the early 2000s made this a high priority and resulted in significant income. This work demonstrated that the laboratory could solve major issues using cutting edge analysis with funding through user-pays and government assistance.

COMMERCIALISED BUSINESS UNIT 2011

The laboratory was aware of the consequences of moving to a full user-pays model. The South Australian Government Laboratory had been broken up, and laboratories throughout the world, especially in the UK, had suffered a loss of approximately one third of the staff in the first year. In 2011, the government decided that the Forensic and Scientific Services laboratory, including the chemical analysis unit, should move to become a commercialised business unit.

COMPETITION

Up until the 1980s, there was very little competition for chemical laboratory services in Brisbane. Some small laboratories existed doing routine chemical analysis of food, minerals, and water. This changed in the 1980s. In 1980, Australian Laboratory Services (ALS) was a small laboratory in Brisbane providing analytical chemistry services for the oil shale and mineral exploration industry.

Australian Laboratory Services was acquired by Campbell Brothers in 1981 and through various acquisitions and expansions became a global laboratory network. Its principal focus was environment and mineral analysis. The presence of a significant commercial global laboratory headquarters in Brisbane presented a significant challenge to a government operated laboratory. When the then Minister for the Environment opened the ALS laboratory and declared that the government would be using it for the analysis of its environmental samples, the laboratory was very concerned about the fate of its work.

A degree of collaboration has developed over the years with ALS and the Queensland Health chemistry laboratories. ALS carried out routine analysis when the number of requests made it profitable to do so. Low volume or complex work was often outsourced by ALS back to Queensland Health chemistry laboratories. A role for the laboratory developed out of this competitive environment. The laboratory reacted to new chemical threats to the community by developing methodology and carrying out high-quality analysis to guide government decision-making. This initial development work was usually followed by several years in which the work grew. Eventually, it reached the stage where the government would decide to put the work out to competitive tender. Due to the overheads inherent in running a government organisation, this work was usually lost to the private sector. Examples of this include the Southeast Queensland Water contract and Corrective Services urine drug screening. On several occasions, the more difficult tasks or the low-volume work was outsourced by the successful tenderer back to the Queensland Health chemistry laboratories. On other occasions, the contract was suspended due to the poor quality of work produced by the successful tenderer, which usually offered the lowest price. It was not uncommon for the successful tenderer to ring the lab and ask how we did the analysis. Fortunately for the laboratory, there was always growth in some area of work that was able to be used to occupy the analysts who previously performed these tasks. Senior officers within Queensland Health were, however, very supportive as they had a series of 'lessons learned' due to outsourcing. For example, even though co-located in the laboratory NRCET sent samples to a university lab rather than use the Queensland Health chemistry laboratories because of price, only to find our results were correct and those of the university were erroneous. Unfortunately, this lesson had to be learned by many of our clients.

An inflexible arrangement relating to personnel made it difficult for the laboratory to change staffing numbers rapidly to meet contractual arrangements. Other factors often stymied the competitiveness of the laboratory. The laboratory could not run loss leaders and then increase the price once competitors had moved from the field. The laboratory information management system was not suited to meet client needs.

ANNUAL REPORTS

For more than 100 years, until about 1995, the lab produced an annual report which was presented to parliament indicating work volume and type, including unusual and significant cases. It was also an opportunity to put on the record shortcomings and deficiencies in the system. On one occasion it was reported that every motorcycle death in Queensland that year was cannabis positive. Subsequent reports created by the larger organisation were more glossy marketing exercises and lacked the detail of previous reports. They served a different purpose. A lot of history has been lost. In the past the media regularly reported the contents of annual reports.

GLUTARALDEHYDE IN ENDOSCOPY CLINICS

In the 1990s and early 2000s glutaraldehyde was used as a high-level disinfectant in endoscopy clinics throughout Queensland. The laboratory purchased a specialised meter for determining glutaraldehyde levels and George Lee, a chemist, moved throughout the State carrying out tests and making recommendations for improved ventilation. On one occasion glutaraldehyde was found in a building not associated with the endoscopy clinic. George rapidly determined that the drain under the endoscopy clinic had overflowed. This liquid then flowed into a gutter which passed the air intakes of the building where the problem was observed. On another occasion, George found that the recycling carbon filters system used to remove the glutaraldehyde from the room air, were saturated. As a result, more glutaraldehyde was coming out of the filter than in the air entering the filter at that time. Eventually another chemical was used for disinfection which *was much safer. Inevitably the time came when George was on the other end of the endoscope, one of the staff recognised him as the "glut man."*

GOVERNMENT ANALYST.

Annual Report.

BRISBANE, October 21.

The annual report of the Government analyst for the year ended June 30, 1932, discloses out of 1865 samples of milk from various parts of Queensland examined by officers of the Government Analyst's Department, during the past year 1500 passed as satisfactory, and 361 failed to pass the test. The analysis showed 4.1 per cent. of the samples were adulterated with water, 7.9 per cent. deficient in fat, 7.5 per cent. were genuine, but below the standard, while 80.4 passed the standard, and 0.2 per cent. were unsuitable for an exact analysis.

The report added that 32 samples of cabbage and six samples of cauliflower contained lead arsenate.

Government Analyst Busy.

A distinct increase in the work of his department for the year ended June 30 was outlined by the Government Analyst (Mr. J. B. Henderson) in the report that the Treasurer (Mr. W. H. Barnes) tabled in the Legislative Assembly yesterday. For the Police Department 208 samples were analysed. "Strychnine has of late years been gradually becoming more prominent in these cases, mostly suicides, because it is now the most readily obtainable poison," Mr. Henderson said. "Sodium fluoride, which is used as a cockroach powder, provided its first recorded fatal case in Queensland." Analyses of milk samples, he said, showed results by no means good, but results that were a marked improvement on those for last year—an average of 9 per cent. of added water, as against an average of 11 per cent. in 1928-29. Ten samples of cabbages received contained lead arsenate ranging from 3.2 to 9.3 grains per cabbage, "showing that this dangerous menace is not yet removed," Mr. Henderson commented. He also deprecated the continued discovery of excess sulphur dioxide in preserved fruit and its use at all in minced meat.

CHAPTER 8: QUALITY SYSTEMS - NATA

Centralised documentation of methods and processes under ISO 9000

(Drawing courtesy of George Lee)

NATA ACCREDITATION

NATA is the National Association of Testing Authorities. Its role is to assure the accuracy of testing results through quality system audits and audits by experts in the field. From a legal viewpoint NATA accreditation is essential. In a legal argument between two labs the Judge will accept the result from the NATA lab.

At a regular audit we were accredited for the analysis of carbon black in polyethylene pipe. We went through the entire audit process, started issuing NATA certificates only to find out that NATA did not record the approval of the method. At the next audit they made us recall all the certificates. It is very disturbing to enter an audit where the first thing said is that we were not signatories. Years later NATA admitted that they were losing track of the accreditations. There was never an apology.

NEW STAFF NEED TO BE CAREFULLY MONITORED

The lab carried out initial urine drug screening with a manual enzyme-based technique. The technique was time-dependent and needed to be carried out methodically to obtain the right answer. A new staff member was trained to perform the analysis including the timing of the various stages. Confirmation of the presence of the drug was carried out by a different technique. A lot of false positives and false negatives were found when comparing the two techniques. The enzyme *technique was reviewed, and it was found the new staff member was preparing batches of samples in advance throwing out the timing of readings. Later an automated system was obtained that removed the operator error and reduced repetitive strain injury to operators.*

GCL acquired NATA accreditation in 1951 for several classes of tests and it has maintained its accreditation ever since. Initially, accreditation was only for the public health chemistry areas as there was no accreditation available for the forensic laboratories.

In 1993, the Queensland Government decided that all businesses it dealt with required an ISO 9000 quality system. As the organisation, QHSS, did not have a quality system, this meant that many of our clients were 'not allowed' to use our services. The Director at the time, Des Connell, was lobbied by Geoff Rynja, Gary Golding and Sharyn Nilsen and decided to pursue a quality systems registration. The laboratory obtained ISO 9002 accreditation two years later and then later ISO 9001 accreditation. In 1998, NATA issued the laboratory with the new forensic accreditation.

The laboratory later gained approval for Therapeutic Goods Administration (TGA) work which was necessary for the analysis of raw materials used in the manufacture of pharmaceuticals. Accreditation for certified reference materials production including alcohol solutions used for calibration of the breathalysers and environmental nutrients was also achieved. The certifications audit had to be delayed or the GCL would have been the first lab in Australia certified for reference materials ahead of the premier reference lab at the National Measurement Institute that had not yet achieved accreditation. Accreditation for proficiency studies in environmental nutrients analysis was obtained mainly due to the efforts of Dan Wruck. His efforts significantly improved the quality of environmental nutrient analysis in Australia and throughout Asia through advisory services, training, and proficiency studies.

Quality systems were originally designed to eliminate a 20 per cent rework that was present in most operations. This gain came at a cost, in some cases estimated to be about 20 per cent of work effort. Thousands of documents generated over the next 15 years required review on a periodic basis. However, accreditation did give the laboratory a level of integrity. It resulted in significant improvements in the documentation of policies, procedures, and methodologies.

Quality systems also changed the work performed by some areas. An unexpected implication for the commercial product testing area of the laboratory was that many government departments began to rely upon the accreditation of the provider to ensure the quality of the product and ceased to test incoming items. The number of textile samples received for testing dropped to almost zero. As a check, three fabrics were tested. Two failed the strength specification. Investigations showed that the white cloth passed but the two dyed cloths failed. The quality certificate provided by the vendor was based upon the undyed white cloth and did not consider the deterioration in strength due to dyeing processes. Despite evidence that the system was not working, no further samples were submitted from the government client, and the laboratory suspended textile testing.

Similarly, before the introduction of quality systems in industry, paint samples used on government buildings regularly failed to meet specifications. Checking of these also ceased. Quality systems allowed the government to procure resources based upon the quality system of the tenderer thus bypassing the laboratory until something went wrong.

STAFFING

In the 1880s Mar had one staff member – a messenger. The estimates for 1959-1960, however, allowed for a complement of 45 professional staff supported by 14 other officers.

The staff numbers grew in the 1980s to approximately 110 professional and support staff. In the 2000s, due mainly to the need to provide more rapid results for the legal system, the number of staff in the chemistry laboratories rose to approximately 142 professional staff and four administrative staff. This number dropped during 2012 due to mandated Government cutbacks. Staff numbers within the chemistry laboratories were barely above 100.

The budget grew from $4.5 million in the early 1990s to $18.3 million in 2012. Earnings from fee-for-service work grew over time from around $1 million in the early 1990s to $6 million in 2012.

Until the 1970s there was only minor use of robotic instruments largely to inject samples into gas chromatographs. The analyst sat by the instrument and recorded the retention times of peaks.

The introduction of computer analysis of chromatograms freed up time for improvements in methodology and the number of analytes quantified in a single run. By the 2010s, hundreds of analytes could be analysed in a single chromatogram with little intervention by the analyst. The digital revolution associated with the interfacing of computers with instruments did not result in less work for staff as we were able to increase the number of services offered and the number of samples that could be analysed through running instruments overnight and over the weekend.

SALARIES AND WORKING CONDITIONS

It is worthwhile to consider the salary side of the laboratory. In the early 1970s, it was difficult to find a gazetted position in the public service that was paid more than a Professional Officer level 4 (PO4) Senior Chemist. Their pay rate was like a back bencher in parliament. This situation changed over time to the detriment of chemists.

The staff could gain promotion from Chemist PO3 to PO4 Senior Chemist through a process of evaluations involving published papers and a thesis on some aspect of their work. Because promotion was related to personal merit, their replacement was usually appointed at the original lower position.

This situation changed in 2010 when evaluation of existing position descriptions and reclassifications within a new industrial award took place. Using this approach, 50 per cent of the staff received a full level increase in their salaries. The staff finally received appropriate and fair compensation for their skills.

The system of job classification meant that there was very little scope for pay increases. Pay levels were often related to the level of the position you reported to. Thus, salary levels depended upon how many levels there were in the department's organisational structure.

Chemists from other areas of government, in smaller departments, were often only two levels below their Director-General. In a large department, chemists could not aspire to higher levels within the organisation, so many left for better pay elsewhere.

CHEMISTRY QUALIFICATIONS

In the period after the Second World War, the sugar industry was growing, and the technical colleges offered a Diploma in Sugar Chemistry to enable returned servicemen to work in quality control at sugar mills. During this time, several chemists at the Laboratory acquired this diploma.

In the 1960s, the qualification for chemists was increased to include a Degree in Chemistry, or a Certificate in Chemistry for the laboratory technicians. For a period, this included the option of a degree accredited by the Royal Australian Chemical Institute (RACI). The universities produced courses which were called chemistry degrees. This specific naming of the discipline made testing the qualifications of applicants a simple matter. In the 1960s and 1970s, a Degree in Chemistry was easily defined; these were generalist degrees preparing the student to work in any area of chemistry.

Once the universities saw the potential of forensic degrees, many students, inspired by the popular forensic TV programs, took up these courses. The feedback from chemistry laboratories was that staff in the forensic chemistry areas needed analytical chemistry degrees so that they would be considered experts in court. Students still flock to the forensic science degrees which has led to a shortage of people with appropriate chemistry qualifications.

EYE IRRITATION IN HOSPITAL

A local hospital's Sterilisation Services Department was having problems with eye irritation. It was determined that the department was using an enzyme detergent for sterilisation, putting it in a sink, and filling it with water, creating an aerosol which irritated the eyes. The solution was to add the enzyme detergent to the sink after it was filled. Chemists at the time experienced a similar problem when they tested enzyme tablets. The first step in our testing process was to grind up the tablets, using a coffee grinder. This produced a fine dust that irritated their eyes. Once the problem was identified the samples were ground up in a fume cupboard.

CHAPTER 9: AMALGAMATION WITH HEALTH PHYSICS

"In nothing do men more nearly approach the gods than in giving health to men."
Cicero 106 - 143 BC

In 1995 it was decided to separate the Radiation Health Division into two groups, one to regulate radiation, and the other to provide analytical testing. They recognised the value of having an independent laboratory carry out the tests for them. Two health physicists, two physics technicians and one administrative officer were transferred to the laboratory at Coopers Plains. For some years they were managed within the Investigative Chemistry group. They brought with them the Queensland Radiation Monitoring Service. The service provided dosimetry badges to government and private sector clients throughout the state. The Health Physics group eventually split off from Investigative Chemistry and became a section in its own right. The dosimetry service ran an in-house software package. Due to software difficulties, it was decided to suspend this work and release staff for other growing areas of health physics work. The resources were reassigned to work involving recycled water analysis. This period was a critical time for government generally, as a long drought had resulted in very low dam capacity. All areas of the laboratory were involved in the testing of recycled sewage and desalinated water. Many analytes were determined including radionuclides. The instrumental resources for recycled water testing in the area grew along with the number of staff. Other areas of work involved clean-up of contaminated radiation sites, including mines and developments on old mineral sand sites. Projects led by Ross Kleinschmidt monitored iodine 131 from medical procedures in sewers and in the waters of Moreton Bay. The sensitivity of the technique is astounding. This provided a useful tracer to monitor flows in the bay. The amalgamation enabled a vast expansion of radiation instrumentation capability.

RADON IN NATURAL GAS

The health physics group performed several consultancies. Whereas most chemistry reports were only a few pages of data, the health physics group produced comparatively long professional reports containing a lot of interpretation, safety advice and regulatory considerations. In some cases, as occurred with a real estate development on an old mineral sand mining site they also produced 3d profile maps of the site showing the lenses of the radioactive mineral sand monazite. In other cases, they tested for radon in natural gas

.

DOSIMETRY SERVICE UNUSUAL HIGH READINGS

The laboratory acquired a radiation dosimetry service with the amalgamation of the health physics group from Queensland Radiation Health. This service sent dosimetry badges to clients throughout the state including Queensland Health facilities and some private practices. At one stage some high readings were obtained from some practices. This is very unusual and was investigated thoroughly. No reason could be found for the high doses. To confuse matters further these high doses tended to occur in the same institutions. It was thought initially that the dosimetry badges may have been exposed to a radiation source during transport to the facility. After much investigation including on-site monitoring, it was discovered that the technician was leaving the stacks of dosimetry badges in the instrument. These badges were processed in the same order each time. The instrument contained a radiation source which was used to calibrate exposure. This radiation source was irradiating some badges at the same level in the stack each time. This meant that similar institutions received badges from the same position in the stack. Hence the coincidence of high readings and institutions. Small seemingly insignificant changes in process can result in dramatic issues.

IT'S NOT RED EMPEROR

Cartons of fish were imported from Asia labelled 'Red Emperor.' The carton contained other types of fish including Unicorn Tusk fish. DNA testing was used to identify the fish.

EXTORTION

A large retail store received an extortion threat claiming that one of their products had been poisoned. The product, consisting of a box of lollies with a white powder added, was found, and analysed. After heating the powder in a flame, there was no change, however there was a very small amount of smoke indicating the presence of a volatile material. X-ray fluorescence analysis showed it to be mainly a silicate material, not

a poison. Extraction of the sample produced a positive for the pesticide propoxur. Later there was another extortion attempt involving the contamination of baby food with caustic soda. When initially offered, the baby would not eat the food. The mother then tried the food, and it tasted very bad, so she called the police. A third case involved the contamination of jam with rat poison. The person was finally caught when he reported a case of contamination. Ten years later there was a similar extortion attempt. Strangely, the person involved in the previous cases, had just been released from prison and just died. Other investigations by the laboratory involved pharmaceuticals laced with strychnine and biscuits containing pesticide.

SORBIC ACID IN CHEESE

Children at a childcare centre developed a rash. Tests by Neil Douglas on cheese they had eaten detected sorbic acid at 1000 times the allowed level.

CHAPTER 10: LABORATORY INFORMATION MANAGEMENT SYSTEM

COMPUTERS IN THE LAB

In the early 1970s, the laboratory purchased its first minicomputer.

The first computer had 4Kb of memory and worked via punch tape and switches. It was used to process data from the recently procured spectrometers. It was also programmed to process data from the automatic gas chromatograph that processed the blood alcohol specimens from suspected drink drivers. Its ability to process large amounts of data saw it applied to generate water analysis reports that were a compilation of results from many chemical techniques. A Chemist, David Grantham wrote a program in BASIC, to generate reports on the assessment of various clays for their potential uses. The long punch tape was routinely fed through the computer into a wastepaper basket until one morning when the cleaning lady threw it out. As a result of the success of these applications and others, the quantity of work for the minicomputer became overwhelming.

In the mid-1970s the laboratory purchased a mainframe computer largely due to the efforts of Keith Deasy. He was chiefly responsible for the introduction of modern instrumentation to the laboratory. Dennis Lee, a chemist, transcribed and refined the program to generate reports on the analysis of water samples. He also wrote a program which interfaced with the gas chromatograph and produced results for the analysis of alcohol in blood for prosecution purposes. When the laboratory moved into a more commercial mode, a program was written which monitored cost codes and prices against each test. This program was run from several terminals throughout the laboratory which interfaced with a central mainframe computer. It quickly became apparent that the laboratory needed a laboratory information management system (LIMS). Firstly Bruce Ames, followed by Michael Ball, a Chemist who moved into information technology, wrote a program that allowed the registration and cost-of-work. This program was called SAMREC. In the early 1990s, this program was upgraded to a program called MARS which was more user-friendly and efficient. Perhaps it was named after the second Government Analyst, Mar.

A very successful program written in-house by Michael Ball was the quality information systems software (QIS). This program was user-friendly, filled many functions and was significant in maintaining the laboratory's quality system. The program was later rolled out to the pathology laboratories. At one point, following a scandal where warnings were ignored, the Director General considered rolling the Opportunity for Quality Improvement module across all of Queensland Health. However, he was replaced before he could implement it. Any person at any level could make improvement suggestions, identify a problem, record fixing a problem or address an audit requirement. The manager had to address the issue in writing. Their responses were audited and if the response was insufficient, they were required to address the issue properly. If they could not, the issue was escalated to a management level which could.

In 1998, all developmental work on MARS ceased by decree, as Queensland Health had purchased a program called AUSLAB which was principally designed with pathology laboratories in mind. It is believed that the committee deciding on the purchase did not recommend this package. It is unclear why it was purchased, but perhaps funding was an issue. To standardise pathology across the state there was a critical need for a common IT system. Standardisation was a struggle. For example, there was not even agreement on the normal range for blood glucose between labs. The chemistry laboratories were required by senior management to adopt this program. Unfortunately, due to specific requirements of chemistry laboratories and clients, this proved difficult and expensive. The impression was created among staff that any constructive criticism of the system was a career-limiting step. This hampered relationship building between the staff and those implementing the program. The cost of making changes to commercial systems is very high and takes time. In a laboratory where the work is continuously changing, turnaround times are critical, analytes suites changing and regulatory requirements varying, any system must be able to be rapidly changed in-house without the high cost of sending the changes back to a vendor. It is not acceptable to have to wait 6 weeks for a client code.

In 2008, a private consultancy developed specifications for a new system and recommended AUSLAB be replaced within two years. Unfortunately, funds were not available at the time to achieve this.

TWO LAB IT SYSTEMS ARE BETTER THAN ONE

Perhaps two IT systems are a good compromise. One satisfies the need for flexibility for low volume non-routine work at minimal in-house change costs. The second, a commercial system for routine work that requires few changes. In a laboratory as varied as this one, it is not reasonable to expect one IT system to satisfy all needs. Contrary to normal practise where the product is designed to meet client needs, IT systems seem to require the client to change practise to fit the IT system.

ODOUR IN ACRYLIC GARMENT SHIPMENT

Acrylic jumpers were imported that had an acrid smell. GCMS identified the presence of a chemical used as a stenching agent in fumigation gas. Stenching agents are foul smelling substances added to gases to ensure gas leaks are easily detected. The container used to import the jumpers was excessively fumigated resulting in a transfer of the stenching agent to the jumpers.

CHAPTER 11: THE LIBRARY

Provided by Geoff Rynja and Trish Murphy

"All men by nature desire to know".

Aristotle 384 – 322 BC

In 1966, the library was in the sub-basement on the way to the director's office. All office staff, including the library staff, were moved from the subbasement to William Street basement level around 1968. This move resulted in a pooling of the office staff. The library was shifted to the Forestry building across the road. It was then moved to the Commissariat building but was brought back to William Street on the first level. In the mid-1970s it was moved to the sub-basement next to the Toxicology laboratory in the old agricultural chemistry laboratories.

The laboratory did not have a librarian, so Lyn Hennessey was taken offline to do the duties of a librarian. She taught herself the principles of the Dewey Decimal System and started cataloguing books and the journal distribution system.

The GCL library was long considered part of the wider Queensland Health library system. John Sutherland was instrumental in the development of the GCL library as an independent resource and introduced one of the first automated library management systems in Queensland Government libraries. John left to take up another position in July 1988. Just before the move to Coopers Plains, Lyn McMillan was appointed as librarian.

In the early 1970s, an arsonist broke into the building and set fire to the library destroying almost 100 years of accumulated journals and scientific books. Many of the books had their backs burned and were taped back together.

Journals were routinely circulated by the library in the 1970s. It could take up to six months for the journals to do the rounds of the laboratory and even longer if they were mislaid. This problem resulted in a restriction on the circulation of journals. In the move to Coopers Plains, specialist library facilities were included in the development. The amalgamation of the LMP bought further library facilities to the building.

The library at Coopers Plains was opened in 1989, with library staff being one of the last tenants to leave William Street. In 1994 there was one typewriter and one computer between the library staff. Requests for document deliveries were typed and then faxed or posted to member libraries, including the British Library. Facilities were extended in 2000 as part of the stage 3 building project.

The library staff were one of the first to use and access the Internet via the collaborative networks with the University of Queensland. SciFinder Scholar was available from the mid- 1990s. Information became much easier to access with databases such as Medline, ABN, Dialog and Pergamon Infoline as well as CDROM content on a local computer.

It was at this time when the scope of work increased hugely, with online literature searches and easy access to public information. At this time, reciprocal agreements with local universities were set into place and have continued into current practises with reciprocal borrowing agreements with government libraries and other Australian libraries via the National Library of Australia.

At various times attempts were made to centralise these in the library collection, however if they were out of the office, staff would often forget that these texts existed. So-called 'bench resources' were essential for training new staff and required to be on hand for quick reference.

The library undertook to identify and catalogue these items, then loan them on a permanent basis to the area. Only some staff took advantage of these specialist texts which once read quickly made them the sectional expert on that subject.

Imelda Ryan became the Library Manager after Lyn McMillan's retirement and built the facilities, staff, and services up to a level befitting a high-quality scientific and research organisation.

ARSENIC POISONING

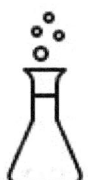

A group of friends had a holiday shack. Several became ill after a weekend at the shack. On the next visit the dog died. Tests on the tank water showed high levels of arsenic. Apparently, the water was drawn from a well near an old sheep dip.

MORE ARSENIC POISONING

A person was admitted to hospital with an unidentified illness. During the medical rounds at the hospital, one of the student doctors suggested the possibility of arsenic poisoning. Arsenic poisoning was later confirmed following a house search, where a vast range of chemicals were discovered behind a reinforced door. Many of these chemicals were the precursors for illicit drugs. There was also a bottle of arsenic trioxide which was half empty. Also present at the site, there were various unsigned versions of the will of the property owner, leaving everything to his cook. Tests on the plates confirmed the presence of arsenic. An associate of the cook died of unknown causes in New South Wales, whilst another was found murdered with his hands cut off.

IT'S SOMETIMES OK TO EAT ARSENIC

With the development of the method for determination of arsenic species in urine one of the volunteer participants was found to have high levels of arsenobetaine. This originated from reef fish the volunteer had eaten the night before. Arsenobetaine is a non-toxic organic derivative of arsenic and occurs naturally in some seafoods. This demonstrates the importance of knowing how arsenic is bound in foods. This is critical in any toxicity assessment. If total arsenic had been quantified without speciation a false impression of the toxicity of the product would occur.

LEAD POISONING OF SHIP PROPELLOR POLISHER

A person was admitted to hospital twice with lead poisoning. He had the classic blue gum symptoms of chronic lead poisoning. His job was polishing propellors from large ships, which he was doing without adequate breathing protection. Investigation of the formula for

Admiralty bronze showed it contained lead. With Doctor Rathus, lab staff visited the works and took samples. This confirmed the source of lead. As usually happens with this type of industrial visit, as soon as the manager disappears from the scene, a union rep or staff member will come up and question the analyst about the results. In the 1970s we could say we have to take the sample back to the laboratory for analysis. Now with onsite testing this situation is more difficult to handle

CHAPTER 12: SPIN OFF LABORATORIES

The trend to consolidate the laboratories to maximise efficiency by minimising overheads has to be balanced against the need for highly specialised facilities. Departments seem willing to fund these specialised facilities provided they remain under their control but are unwilling to pay for the same service through a central laboratory.

In the late 1970s, following several explosions in coal mines, a decision was taken to set up the safety in mines testing and research station (SIMTARS) at Redbank. The GCL had previously undertaken work in mine safety at the direction of the Mine's Inspectorate. This work included sampling dust and gases in mines throughout Queensland. These services and several staff members were reallocated to the new facility at Redbank where the laboratory continues to this day. This laboratory has been able to focus specifically on mining work. However, the need for fee-for-service income and the synergy with occupational health and safety analysis, has resulted in diversifying beyond just mine safety work. The work of this laboratory has led to a significant drop in the number of fatal mine explosions.

In 1991, with the increased focus on workplace health and safety, a decision was taken to allocate laboratory staff previously occupied in occupational hygiene analysis at the GCL, to the Department of Workplace Health and Safety, the concept being that they would set up a specialised laboratory to continue this work. However, the new department priorities were more focused on non-chemical areas of occupational health and safety, as large numbers of injuries and deaths occurred in the building industry. Many of the transferred laboratory staff took up positions as advisers and moved into the inspectorate and managerial fields. Years later, black lung and silicosis re-emerged as an issue. Dust monitoring had been a significant expertise area of the occupational hygiene analysis section of the GCL, but this service was lost when the section was moved to another department.

LEGISLATED LIMIT OF ZERO LEAD IN PAINT

At one stage, the government decided to implement a limit of 'nil' lead in paint. The increase in sensitivity of analytical instruments in the latter half of the 1970s meant lead could be detected in most items analysed.

It took considerable effort to have this limit raised to a logical and safe level. With the renovation of old Queenslander houses in the early 2000s, levels of lead in paint flakes were still found to be up to 30 per cent.

BOMBING OF THE SS ARAMAC –

The Argus (Melbourne) 3 August 1893

THE EXPLOSION ON THE ARAMAC.

REPORT BY THE GOVERNMENT ANALYST.

BRISBANE, WEDNESDAY.

The fragments of the infernal machine, which were collected on board the Aramac after the explosion, have been analysed by the Government analyst. He states that the substance adhering to the hair, and also that adhering to the piece of tin, is the explosive generally known as gelatine dynamite. Smaller pieces of dynamite are distributed throughout the other samples. Gelatine dynamite, the analyst states, consists essentially of a mixture of nitro-glycerine and gun-cotton collodion.

SYDNEY, WEDNESDAY.

Details have been received in Sydney of the view the Queensland police take of the dynamite explosion on the steamer Aramac. Their opinion is that the explosion was arranged in Sydney, and the information given by passengers by the steamer seems to bear out this view. Some little time before the Aramac left the wharf at Sydney a man was observed by some passengers to leave the forecabin and pass hastily up the companion-way. This man came from the port side of the vessel, where no passengers were accommodated. He carried a parcel, and his actions were so suspicious that he was watched. He was noticed to walk up to another person, who shook him by the arm as if congratulating him on what had been done. Nothing further was thought about this occurrence until now that an investigation is being conducted, but the police believe that this incident was connected with the outrage.

CHAPTER 13: CHEMICAL EMERGENCY RESPONSE

The Director in the 1980s, Trevor Beckmann, took a personal interest in this area and was often called out to attend emergencies. These events were rare although occasionally involved significant hazards such as the assessment and destruction of mustard gas washed up on the shores of Moreton Island.

Due to the rare nature of these events, they were dealt with on an ad hoc basis. In 1993 the Department of Emergency Services decided to set up a specialised unit to respond to these emergencies. The GCL provided four chemists (Ron Biltoft, George Lee, Ross Sadler, and Gary Golding) to assist with this operation. They were trained to wear and decontaminate fully encapsulated suits.

> The Queensland Fire and Emergency Services was able to allocate considerable resources to set up on-site mobile analytical facilities to enable assessments to be made at the scene.

ODOUR IN CINEMA

Laboratory staff were called out to a cinema theatre complex in Fortitude Valley, following reports of a strong odour of chemicals. The theatre was evacuated, and tests rapidly identified the odour as xylene and the source determined to be some renovations in another part of the theatre which included the toilets. The exhaust from this area was vented to the roof, adjacent to the air conditioning intake which recycled the xylene into the main theatre.

BRISBANE EXPLOSION

EVIDENCE OF ANALYST.

5LB OF EXPLOSIVES USED.

BRISBANE, Tuesday.

Further evidence was heard in the Police Court to-day in the case in which Albert Orchard, miner, was charged with placing gelignite in the offices of the Criminal Investigation Department on the night of August 13.

John Brownlie Henderson, Government analyst for Queensland, stated that he came to the conclusion that the damage was caused by high explosive, and that a quantity not less than 5lb in weight had been used. Gelignite thrown through a window in George-street could have caused the damage.

Detective Acting Sergeant Henderson stated that John Joseph Cahill identified defendant in a line-up as being one of two men he had seen standing near the office of the Commissioner for Police about 11 o'clock on the night of the explosion.

Defendant was further remanded.

BOMBING ON GEORGE STREET

Sydney Morning Herald, 31 Aug 1927

Throughout its history the laboratory has been involved in the investigation of explosions. These included explosions at the central post office, coal mine explosions, bombings, a gas explosion in Queen Street, clandestine laboratory explosions and plane crashes involving explosives.

ACID CLOUD OVER FACTORY

In one call out, George Lee attended a factory near the Queensland Police Academy at Oxley. When he arrived at the gate of the factory, a heavy acrid mist was present. The factory had been manufacturing a chloroparaffin glue for use in an insulation product. Unfortunately, a thermostat on the reactor malfunctioned and caused the mix to over-heat, resulting in the production of hydrochloric acid gas. It was a very cold evening, and the gas dissolved in the dew on the rafters of the factory. This absorbed hydrochloric acid, dripped down on the concrete, causing lines of corrosion to become visible on the floor. George's extensive experience in the chemical industry in the UK quickly determined the cause.

ARE THESE PRAWNS AUSTRALIAN?

The acquisition of Isotope Ratio Mass Spectrometry instruments and the employment in 2012 of Dr Jim Carter, a world authority in the field, enabled the laboratory to test the provenance of food. One project looked at the difference between Australian farm grown prawns and those imported from overseas. The resulting technique was used to help prosecute a vendor who was selling Malaysian prawns as locally sourced from Moreton Bay. The technique has also been used to examine other consumables, including honey, coffee, beer, cider and to examine illicit drugs.

RED TAPE DOES EXIST

Common terminology in the public service is that requests to government were tangled up in red tape. People are probably unaware that red tape does exist (see photo). This red-tape was used to seal up samples sent to the laboratory by environmental health officers. They also used red wax seals to protect the legal integrity of the sample.

CHAPTER 14: AUSTRALIAN FUTURE FORENSICS INNOVATION NETWORK- (AFFIN)

Provided by Graeme White and Dennis Burns

Planning, establishment and securing of funding support for the Australian Future Forensics Innovation Network (AFFIN) was initiated and principally led by Graeme White, then Principal Adviser, Forensic Sciences, QHFSS and Dennis Burns, then Associate Professor and Director of Forensic Science at Brisbane's Griffith University. QHFSS was involved with full support from the Senior Director, Greg Shaw.

AFFIN, supported by funding over the period 2007–2012, was Australia's first national multi-organisational network that carried out coordinated, collaborative, industry-focused, forensic science research, development, and innovation. AFFIN's focus during that period, across seven research and development projects, was development of new and innovative 'in-field' forensic analytical technologies and capabilities. AFFIN's activities were also to include developing systems for knowledge and technology transfer within the forensic industry, supporting operational needs, thereby contributing to upgrading of skills and technologies. AFFIN was also required to seek out opportunities for commercialisation of its project outputs.

Funding support for AFFIN was secured in two stages. Stage 1 involved the award in 2007 of a Partnerships-Alliances Facilitation Program (PAFP) grant by the Queensland State Government's then Department of Tourism, Regional Development, and Industry (DTRDI). The collaborating partner organisations contributed cash and in-kind funding to support the PAFP grant.

PAFP grant funding was used to:

- assist in the development of a national framework for coordinated, collaborative, forensic research targeted to meet industry needs
- engage the services of Dr Patrick Silvey and his team from VenturePro Pty Ltd to assist in developing a business case for progressing the AFFIN initiative
- support a national planning workshop with input from a range of forensic stakeholder organisations across Australia

Stage 2 involved the award in 2009 of a National and International Research Alliances Program (NIRAP) grant by the Queensland State Government's then Department of Employment, Economic Development, and Innovation (DEEDI). The collaborating partner organisations contributed cash and in-kind funding to this NIRAP grant.

NIRAP grant funding was used to establish AFFIN and to support its seven 'inaugural' forensic science research and development projects.

Establishing professional links, advice, and support from Maree Storer, then Principal Project Officer, Research and Development Services from the then Department of Employment and Economic Development (DEED) and Phil Abernethy then Business Innovation Services from the then Department of Employment and Economic Development also played important roles in establishing AFFIN.

The Queensland Government maintained close ties with AFFIN throughout 2007– 2012 and was proactive in strategies to assist in AFFIN's development to provide sustainable, long-term support for forensic science research and development with outputs that would benefit industry.

Initially, AFFIN was established as an unincorporated joint venture of 20 collaborating partner organisations including Australia New Zealand Policing Advisory Agency (ANZPAA) National Institute of Forensic Science (NIFS); state forensic laboratory service providers, QHFSS, Forensic Science South Australia (FSSA), the Victorian Institute of Forensic Medicine (VIFM), National Forensic Pathology Service New Zealand (NFPS NZ); state police services (AFP, QPS, NSW Police); and universities (Bond University, Griffith University, Edith Cowan University, Flinders University, Queensland University of Technology,

University of Queensland (UQ), University of Tasmania (UTAS), and University of Technology Sydney (UTS)). Bringing all these groups together for a single purpose was a considerable achievement by Graeme White and Dennis Burns, given the competitive nature of those involved.

In accordance with NIRAP grant milestone deliverables AFFIN was to be incorporated and NIRAP grant funding was to be novated (transferred) from the then custodian of the funds (Griffith University as NIRAP grant recipient host) to AFFIN Ltd.

This incorporation and novation were to provide improved flexibility and greater efficiencies through an independent AFFIN to maximise potential of achieving growth and development objectives. AFFIN would have been an independent entity free from possible undue influence from any collaborating university partner organisation. The independence of AFFIN was not only consistent with a major milestone deliverable in the NIRAP grant application but was also a key outcome of the national planning workshop. Unfortunately, this novation of grant funds never happened despite best efforts and so the sustainability of AFFIN beyond the period of the NIRAP grant was not guaranteed. This was undoubtedly a significant lost opportunity for strengthening industry- focused forensic science research and development in Australia.

For an example of AFFIN's key successes, consider the outcomes of one of AFFIN's projects: 'Project 7–

Evaluation of illicit drug use patterns through wastewater analysis. Led by Professor Jochen Mueller, the project parties were UQ, FSS, AFP, Bond University and UTAS. In September 2016, Professor Mueller emailed contributors to the project and stated, 'The project which started with some funding from AFFIN has become a massive success story." Prof Mueller went on to list outcomes including details of 25 research publications in peer-reviewed scientific journals and subsequently awarded grants, built on the outcomes of the AFFIN- funded project 7.

COURT CAN HAVE ITS FUNNY MOMENTS

At one stage during his career the author gave in person evidence in court 19 times in a six-month period.

The author was in the Supreme Court and there was a glass of water and a jug on the dock in front of him. The evidence being completed, the judge said, 'the witness may step down.' he swung around on the swivel chair, his knee hit the dock, and the glass and jug rocked but fortunately did not spill. The judge said, "'I said step down, not knock it down". This lightened the situation somewhat.

In another case, involving a soil match the analyst was called to give evidence at 12.30 pm, with his return flight scheduled to leave Cairns at 1.15 pm. The evidence was completed by 12.40 pm. The judge then started asking clarifying questions including, 'What is soil?' Being a little bit stressed the only answer that came to mind was 'the friable rock particles and decayed organic matter covering the dry surface areas of the planet earth...' probably not the best answer. At 12.50 pm he left the courtroom and got into a police car to the airport. After taking a run down the airbridge to the hostess holding the door open, he boarded the plane back to Brisbane.

The Act covering blood alcohols required the analyst to perform a test (singular). One defence barrister regularly accused the analyst of carrying out three tests because the analysis was conducted in triplicate and thus did not fulfil the requirements of the Act. Once the analyst stated that the one test involved doing the analysis three times and averaging the results, the barrister would say '"you got three different results." On one occasion he tried this approach, and the analyst was able to say 'no, the three results were identical, in each case 180mg/100ml'. Sometimes you get lucky.

On another occasion the barrister was working through a book of questions to ask in blood alcohol cases. He claimed that the *analyst did not perform the test, the instrument did the test, thus the Act was not followed. Fortunately, the analyst had read a report on a judgement from the UK where the Lord Chief Justice had ruled that the analyst uses the instrument much as a carpenter uses a hammer to drive a nail. The analyst repeated this statement mentioning the hammer and raised a smile from the barrister.*

MORE COURT FUN

In another case the analyst flew to Cairns and upon arrival at the Cairns Airport was told the accused pleaded guilty and the police had booked him a return flight home. The hostess did a double take as he re-entered the plane. In the mid-1980s, visiting analysts were booked into a motel near the Cairns railway station at a reduced rate. This motel was later mentioned in one of Matthew Condon's books on the pre-Fitzgerald era Queensland Police Force, as being owned by some undesirable elements.

One analyst, or how he told it, took the opportunity when in Cairns for a court case to purchase fresh seafood to bring back to Brisbane. On this occasion, he purchased some live mud crabs. On boarding the plane, he left them with the crew for safekeeping during the flight. The hostess couldn't resist the temptation. Upon landing she announced over the PA system "Will the gentleman who gave me the crabs in Cairns please come down to the front of the plane!".

HOW DO YOU KNOW IT'S A LOAF OF BREAD?

In the days when baking bread on Sundays was illegal, the analyst was required to go to court to provide evidence. The analyst was asked if he did a test for sugar content, to which he advised he had not. The defence then claimed in the absence of a sugar test, the loaf could have been a cake, which could be legally baked on Sundays. A question along the lines 'How did you know it was a ....' was a trap for new analysts. Usually, the tests assume the nature of the sample and focused upon a component. The same question has been asked in oil spill cases and soil matches.

DESTRUCTION OF CONTROLLED DRUGS

Health agencies including pharmacies, nursing homes, the flying doctors and hospitals regularly sent in controlled drugs for destruction by a State Analyst in accordance with the requirements of the Health Act. The estimated street value of the destroyed drugs was about $30 million per year. Charles Chooi returned from retirement and carried out the counting checks for 12 years until his 77th birthday. It was important to check the contents of every package because there had been instances where some of these drugs were diverted for illicit sale. The laboratories' tightly controlled system became important in these situations. Drugs were sent in by registered mail to provide a tracking mechanism. On one occasion a package went missing after being received at the laboratory. Extensive searches were unsuccessful. About six months later the package was found during the servicing of the lift. Unnoticed, it had slipped off the trolley and down the crack between the lift door and the building's floor, and into the lift well.

CHAPTER 15: AUSTRALIA PACIFIC FOOD ANALYSIS NETWORK AND TRAINING

Provided by Pieter Scheelings

The process of becoming a training organisation commenced with the Asia-Pacific Food Analysis Network (APFAN).

In the late 1990s the laboratory's expertise had increased to such a level and its facilities were of such a high standard that it became a venue for training laboratory chemists from other nations. This process was assisted by the construction of a world-class auditorium for the provision of seminars instigated by the Director Michael Moore.

The genesis of APFAN was during the 1989 Brisbane Conference of the Royal Australia Chemical Institute (RACI) and the third Federation of Asian Chemical Societies (FACS) congress. Several food scientists from developing countries, who attended the events, identified gaps in capability. These included the lack of collaborative opportunities, dated instrumentation and analytical methods, inadequate laboratory facilities, limited reliable food composition analysis and nutritional data for local food supplies. Accordingly, a group of these food scientists decided to form a regional network of food analysts, to exchange methodologies, information and explore training opportunities across the Asia–Pacific region.

A committee of key representatives from interested countries was formed to promote food safety and good nutrition. Dr Howard Bradbury, emeritus professor at the Australian National University offered to take the leading role of APFAN Coordinator, while Graham Craven, the representative from QHSS, offered the laboratory facilities at Coopers Plains as a training location. The network was designed to support food analysts with the motto 'food analysis is vital to food quality and safety.'

As APFAN membership was free, it had 500 registered members including representatives from the Middle East and Africa and included a regular newsletter over several years. 130 members participated in the workshops at QHSS, and many returned home to become drivers of analytical quality in their own country. Some trainees had travel and accommodation costs funded by APFAN through generous donations from agencies such as The Crawford Fund, Australian Centre

for International Agricultural Research (ACIAR), AusAID and FACS. Dr Pieter Scheelings took on the role of coordinator from Howard Bradbury in 2000 which in turn was passed onto Professor Wirarno from Indonesia in 2015.

Over the ensuing decades, APFAN hosted seven practical workshops at Coopers Plains, organised five analytical conferences within the Asia–Pacific region and supported several other training programs and conference travel awards.

The APFAN training programs benefited considerably from the willingness of QHSS management to make staff and facilities available during the training. Likewise, many staff were able to develop their own presentation and training skills as part of the APFAN programs as well as other postgraduate training programs which QHSS undertook in collaboration with EnTOX.

Chemists from overseas worked at the laboratory for periods of up to three months. They came from Indonesia, Thailand, South America, Fiji, and Seychelles. The laboratory also hosted scientists from Papua New Guinea and mainland China to obtain experience in quality systems.

Each year, 20–30 chemists from across the world visited the laboratory for training in food analysis. As part of this process, they were required to analyse a reference material which could be taken back to their own country to act as a quality control. Many of these laboratories had very little funding and certified reference materials were often beyond their financial capabilities. Local staff were surprised to learn of the conditions under which these people worked. In one case they were allowed only a very limited number of latex gloves per month. They often took home copies of our validated methods. These workshops usually lasted about two weeks.

DESTROYING CONTROLLED DRUGS

State analysts had an assigned duty under the Health Act to destroy controlled drugs. Periodically when sufficient drugs were obtained, the drugs were loaded into a car and driven, without police escort to the Willawong incinerator and burned. The decision to burn was made on an ad hoc basis by the supervisor so that nobody could predict when a burn would occur. Arranging an escort meant that many people knew of the transfer, thus increasing the risk. In dramatic contrast, a heavily armed police escort was provided for the Commonwealth Police drug burns. One day when leaving the grounds of the lab a passer-by noting the cannabis in the back of the van said "don't burn it all in the one place". Police escorts later became normal practice.

MISUSE OF CONTROLLED DRUGS

Special care is needed when dealing with controlled drugs. Careful recruitment and monitoring of the process is essential. The misappropriation of controlled drugs is a known problem in hospitals. The laboratory routinely received medicines suspected of being tampered with. These included plastic ampoules with needle holes. Random drug screening was considered but when this was suggested to the Department's Human Resources division, this was not allowed, although it was common on mining work sites in Queensland.

A staff member was recruited to destroy the drugs. He was overheard at a restaurant boasting to his friends that he could get large quantities of drugs. His temporary contract was not renewed. The situations led to a serious review and tightening of the control of the processes around the destruction of drugs. At an ISO 9000 quality audit the assessor said it was the best controlled process they had encountered.

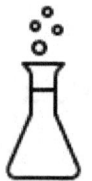 BABIES AND BARBITURATES

A once popular medicine for babies with colic was based on phenobarbitone. It was not a commercial product but made up by the pharmacist. On one occasion a baby slept for two days following a dose of this medicine. The pharmacist had used pentobarbitone which is much more powerful. On another *occasion the pharmacist used the sodium salt not the free phenobarbituric acid. The British Pharmacopoeia (BP) had a method of analysis which assumed the barbiturate was present as the free acid. The analyst obtained low results which did not explain the symptoms. Reanalysis following acidification yielded a high result. An investigation followed which led to a letter to the BP requesting the method be amended. It is worth noting that although many BP methods have been superseded, one analyst in court defending his modern instrumental method eventually told the court that the sample had also been analysed by the BP method. The barrister quickly stopped questioning.*

 ALWAYS RUN YOUR OWN STANDARDS

A batch of paracetamol tablets was being checked by spectrophotometry. The literature value of the absorptivity of a 1% solution of the drug was used instead of a standard. It was taken from a reputable text. Fortunately, the analyst, prior to reporting the failed result, ran a standard, and discovered the error.

CHAPTER 16: AUSTRALIAN AND INTERNATIONAL STANDARDS COMMITTEES

The laboratory was instrumental in the development of Australian (AS) and International Standards (ISO) involving chemical analysis.

Henry Olszowy followed Paul Geoghegan on the committees that produced the Australian Standard for analysis by atomic absorption and inductively coupled plasma spectrometry. Henry was Chairperson of the Committee on Spectroscopic Analysis CH/16 (revision of AS 2134, AS 3641, developing AS4873.1 and AS 4873.2). He was a committee member, of Standards Australia, Major and Minor Components in Coal Ash by XRF Analysis and Aqua Regia Digest of Soils for Heavy Metals Determination by ICP. Ron Biltoft served on the Standards Australia committees on plastics in contact with food. Geoff Rynja was an active member of Standards Australia Committee CH/22 'Examination of Wastes'. It looked at developing the Australian equivalent to the USEPA's 'Toxicity Characteristic Leaching Procedure' (TCLP) for the examination of wastes from contaminated sites. Alan Webb and Gary Golding were on the toy safety committee CS18 of Standards Australia. Gary was on this committee for 21 years during which time they released a series of standards on toy safety, including one for chemistry sets. He saw the need for an international standard which screened the total amount of toxic metals in the surface coatings on toys. The main standard tested for leachable metals. Through Standards Australia he initiated action to commence its development. This new international standard, ISO 8124-5:2015, (Safety of toys — Part 5: Determination of total concentration of certain elements in toys ISO 8124-5:2015 specifies methods of sampling and digestion prior to analysis of the total concentration of the elements, antimony, arsenic, barium, cadmium, chromium, lead, mercury, and selenium from toy materials and from parts of toys) provided a screening technique to avoid the more expensive leachable metal test. Henry Olszowy was instrumental in developing this standard as the main analytical expert on the International Standards Organisation committee. This International standard would be the highest-level standard initiated and developed by the laboratory staff.

International Involvement

APFAN built a network of connections for the laboratory throughout the world and led to the initiation of international collaborative research.

One staff member Tatiana Komarova built collaborations with scientists in Russia, developing passive sampling devices for environmental monitoring. Pieter Scheelings gained funding through the United Nations to travel to Mozambique and Vietnam to provide training in food analysis. Mary Hodge travelled to Indonesia to provide courses on pesticide residue analysis. Henry Olszowy, Geoff

Eaglesham and others travelled to China to provide lectures. The laboratory also developed a world-class environmental water, low-level nutrients laboratory, led by Dan Wruck. Through his efforts, particularly collaborative trials, the standard of environmental nutrients analysis in Queensland and Australia was improved considerably. He later travelled to various parts of Southeast Asia and China giving lectures on nutrient analysis. He also sat on international committees where his expertise was greatly valued. Other staff ran courses in blue-green algae identification. With the drought of the 2000s, blue-green algae flourished in water storage facilities, and it fell to the laboratory to develop methods of assessing the significance of the blooms. Geoff Eaglesham developed methods to quantify the level of algal toxins in the water. These results fed into decisions to close water storage facilities (dams) until the levels were acceptable.

 BRISBANE FLOOD 2011

This flood brought the city to a standstill. There were several deaths, and a lot of damage associated with the floods. One unforeseen outcome was that there were fewer drug samples submitted to the laboratory due to diversion of police resources elsewhere and the inability for drug dealers to get around. There were also fewer autopsies performed because there were fewer road accidents and drug overdoses. This allowed us to catch up on the backlog.

ASIAN TRADITIONAL MEDICINES

In collaboration with Queensland Health environmental health officers, a selection of Asian traditional medicines was tested for heavy metals and scheduled drugs. One contained high levels of mercury, several contained scheduled drugs including one containing a Schedule 9 prohibited substance.

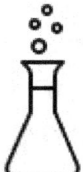

ATTEMPTED RAPE AND FIBRE TRANSFER

Geoff Rynja and Gary Golding developed the capability to perform forensic fibre testing on monofilaments transferred during a crime. This included the extraction of dyes from small quantities of fibres in capillary tubes for thin layer chromatography comparison.

The police received a complaint from a young woman regarding an attempted rape in her bedroom. The evidence included a pen with a suspect's name engraved. It was found on top of the clothing that the girl had removed before getting into bed. There were numerous fibres matching the girl's bedspread on the defendants clothing.

BAD SMELL IN HOUSE

A house holder complained about a bad smell in the kitchen. It seemed to be coming from the kitchen cupboard. The cupboard had been lined with old vinyl floor covering. The smell started soon after the fumigation of the house for West Indian Dry Wood termites with methyl bromide. Chemist George Lee was aware that a reaction between the methyl bromide and the vinyl floor covering could cause the problem. The removal of the vinyl eliminated the problem. This points out the value of the long term accumulation of tacit knowledge. Tacit knowledge is knowledge you don't know you have until something activates it.

CHAPTER 17: PROFESSIONAL BODIES
The Royal Australian Chemical Institute

Membership of professional bodies is particularly important for continued professional development. It also provides excellent networking opportunities for career advancement and problem solving. RACI is the body representing chemists in Australia and throughout its history, chemistry laboratory managers and staff have supported the activities of the RACI in Queensland.

Management provided time off to organise seminars and provided facilities, free of charge, for seminars and conferences.

Many senior staff members from the chemistry laboratory have become fellows of the RACI, and others have gone on to become Queensland Branch Presidents. When the health practitioners scale was instituted, part of the generic requirements for more senior staff was involvement with professional bodies. This demonstrated the commitment of the department to these bodies.

The RACI provides opportunities for chemists to gain leadership and project management skills through managing events from local seminars through to international conferences.

For many years, one of the requirements for chemists to be employed by the GCL was a degree in chemistry or equivalent as recognised by RACI. In recent years this later statement has been dropped. The removal of this criteria may be justified, given that the laboratory requires staff to be proficient in analytical chemistry, not just chemistry generally, which has many specialities.

Many chemistry staff who worked in the areas of forensic toxicology and forensic chemistry were foundation members of the Australian and New Zealand Forensic Science Society.

Clandestine Laboratory Investigating Chemists' Association Peter Vallely was president of this USA Association.

When the laboratory acquired the capacity to measure volatile organic compounds in air, several staff joined the Clean Air Society. Henry Olszowy was a committee member, Qld Branch, Australian X-Ray Analytical Association.

ANALYSIS OF SOFT DRINKS

The laboratory routinely analysed soft drinks for conformity with various regulations. This included caffeine content in cola, fruit juice content in some types of soft drinks and identification of colouring dyes. At one point the manufacturers wanted to put caffeine in all soft drinks e.g, lemonade. The Health Department stopped this occurring at the time but the rise of energy drinks containing caffeine overcame this embargo. On another occasion, a business took excess pear juice from a canning plant, purified it to a clear, odourless, sugary liquid and proposed to use it to replace the water used in soft drinks. They wanted to brand the resultant soft drink '100 per cent fruit juice'.

SCHOOL CHEMICALS

The laboratory received surplus chemicals from public schools for destruction. Unfortunately, the Brisbane council chemical disposal site at Willawong closed and for a period there was no avenue to dispose of the chemicals. A second-hand chemical vendor took some of the chemicals for resale, which were later discovered by our oncall chemists at a clandestine amphetamine laboratory. The laboratory assisted the Education Department to prepare a booklet suggesting methods of disposal of chemicals. This was used for many years. Unfortunately, the Education Department then used private consultants to devise a method that involved mixing different chemicals together. The number of possible chemical interactions increased exponentially with the number of chemicals mixed together.

The new process resulted in the lab being called out to an incident at a school where incompatible chemicals were mixed. The new mixing process was abandoned upon advice from the laboratory.

POWDER IN POLICEWOMEN'S SHOWER

A white powder was found in considerable quantities on the ceiling of the police academy shower. Analysis of the material showed that it was strontium carbonate. It had preferentially leached out of the concrete.

TENDER FOR WOOLLEN BLANKETS 1980S

A local company produced woolen blankets intended for a government hospital contract. These woolen blankets were tested and did not meet the specification required by the tender. A request was made by the Premier's Department for all information relating to these tests to be sent to their office for examination by a third party. Tenders were very political; the unsuccessful party always complained.

CHAPTER 18: WORKPLACE HEALTH AND SAFETY 1998-2013

Provided by Robyn Mackenzie

The Workplace Health and Safety Unit was formally established in 1998 with the appointment of Claire Moore and Lynda Devitt to the roles of Workplace Health and Safety Officer. Until this time health and safety activities had been carried out by laboratory staff on a voluntary basis, however a new and more demanding legislative and business environment provided the impetus for increased health and safety commitment and resourcing.

Driving forces included:

- The WHS Act 1995 and Regulation which set out increased management obligations for the health and safety of staff.
- ISO 9000:2000 Quality Management Systems. To achieve accreditation, the organisation needed to have a documented quality management system, including compliance with relevant safety standards (AS 4804 Safety Management Systems: AS2243 Laboratory Safety Standards).
- A major building project at the Coopers Plains site (Stage 3) which significantly increased the size, complexity, and employee population.
- Increasing client and stakeholder expectations

- The merger with Qld Health Pathology and Scientific Services (QHPSS), and ongoing business / quality / performance requirements.

The establishment of a permanent health and safety unit was a turning point for the organisation and facilitated the development of a laboratory-specific safety management system, including consultative committees, hazard identification, incident investigation, emergency response, executive reporting, and staff training programmes. A mammoth project was undertaken within a short timeframe by Claire Moore and Lynda Devitt.

In addition to the work undertaken by the WHS Unit, staff were supported with rehabilitation and return to work services by Sadet Davis who had relocated to the Coopers Plains complex following amalgamation of the GCL (Government Chemical Laboratory) and LMP (Laboratory of Microbiology and Pathology). Sadet was also instrumental in introducing infection control and biological safety procedures across the campus through the implementation of vaccination and biological monitoring programmes.

Further consolidation and expansion of safety services and systems occurred in the years 2005 - 2013 in response to a Queensland Health safety system audit which identified an urgent need for increased funding and resourcing across all health districts. These findings were reinforced around the same time by the Ministerial Taskforce Review on the Role and Function of Forensic and Scientific Services 2005. The review recognised a high level of risk within the work environment and mandated the maintenance of a dedicated, onsite WHS unit to manage and minimise risks associated with counter terrorism samples. As a direct result of the Queensland Health safety audit, additional staff and funding were allocated. In 2006 Robyn Mackenzie was appointed to the position of WHS Manager. A short time later Sadet Davis joined the team permanently as WHS Officer, along with Lynda Devitt who had continued in the role from 1998. The WHS Team also worked in partnership with Jack Cossart, QHPSS Dangerous Goods Safety Management Officer, to manage and minimise risk associated with approximately 4000 chemicals stored and handled onsite.

Between 2010-2012 Jessica Dixon, Megan Tilley, Rachel Myska, and Nicole D'Arcy also joined the team, contributing their professional, scientific, and administrative expertise to the work and reputation of the unit.

This was a time of rapidly changing technology; political and business priorities and many important milestones were achieved along the way. These provide a chronicle of issues and responses during the period 2005-2013:

- Centralisation of WHS services and activities across the work site and remote work areas to units to achieve consistency and ensure site compliance with legislation and regulatory requirements.

- Compliance with the Dangerous Goods Safety Management Act and Regulation 2005.
- Rollout and maintenance of the chemical management system Chem Alert
- Implementation of the online QFRS Fire and Evacuation Program in line with the Building Fire Safety Regulation 2008.
- Strengthening of workplace consultation, information, and staff education
- Implementation of health monitoring and surveillance programmes for at risk staff (chemical; biological)
- Infection control, vaccination and biological monitoring programmes aligned with Queensland Health and NH&MRC standards.
- Review and improvement of RRTW services to injured staff.
- Implementation of a Healthy Lifestyles Program to facilitate physical and psychological wellbeing in the workplace.
- Undertake hazard identification inspections and audits across all layers of the organisation to identify issues, prevent injury and ensure staff safety.
- Respond to, and investigate safety incidents to identify contributing factors, initiate change and facilitate continuous improvement. Notable examples include an ultra-light plane crash within the FSS grounds, chemical incidents (related to incompatible substances / exposure / spills), biological exposures and a legionella outbreak follow-up.

CHAPTER 19: REVIEW
"Those who cannot remember the past are condemned to repeat it".
George Santayana 1863 – 1952

Throughout the past 150 years the reason for the chemistry laboratories existence has often been ill-defined by government other than for a few Acts which require a State Analyst to carry out a particular analysis. Despite this, the work done in chemistry laboratories fits with the broad objectives of government: a secure society, a healthy society, an educated society, and economic development.

For this reason, analyses being requested by clients over the past 150 years has remained remarkably unchanged. As Karl Staiger did in 1873, the laboratory still carries out water, food, toxicological, environmental, and forensic analysis for the police service and other government departments.

The laboratory has always been called upon for new or non-routine chemical analyses. Many of the analyses carried out are low-volume, unique, or highly complex, subject to unexpected analytical interferences, require value added opinions and interpretation, and hence are expensive to perform and probably not profitable for the private sector.

New work-types move through a series of steps. Initially a few samples require analysis for a non-routine analyte. These results may highlight a problem, which can cause action to be initiated by government. Over time the work grows as regulations are enforced, and routine monitoring begins. In the case of organisations that are charged for the analyses, they eventually reach a stage where they seek cheaper analyses elsewhere. This is usually by competitive tendering, because at this stage the work has reached such a volume that it is commercially viable. There have been instances where the work won through the lowest price tenders by private laboratories is returned later to the laboratory because of quality issues. Indeed, private laboratories have been known to contact the GCL requesting methods or to outsource the more complex work back to the laboratory.

The laboratory is called upon to carry out urgent analyses to provide data for government decision-making. It has been able to provide this service because it is well-equipped, has a flexible workforce (including generalist chemists), an ability to interpret results, and has a system up and running that can be altered at a moment's notice to accommodate a government emergency.

The laboratory has always been at the forefront of adopting new instrumentation. There have been dramatic changes in technology. Detection limits have decreased with each new instrumental revolution opening new areas of investigation. Increased community concerns about their exposure to chemicals has facilitated the need for methodologies that can detect extremely low levels of materials.

Fortunately, the laboratory has always encouraged staff to stay up to-date with the latest technologies and be able to appreciate the application of this technology in the real world. This has happened because the field of chemistry is of such a broad scope and complexity, that one person could not foresee all potential applications.

All chemists must take a leading role in the implementation of new analyses. Distributed leadership has been, and should continue to be, used to empower individuals at all levels to exert positive influence over change and development. Above all, the management of scientists requires unique capabilities. They need to encourage staff, ensure their continual development, and support new initiatives. Most importantly, managers need to find mechanisms to attract and keep the best graduates in analytical chemistry.

All scientific projects are a compromise of time, quality, and cost. To maintain quality, appropriate resources need to be provided or timelines are compromised. Getting analytical results quick, cheap, and correct is a difficult task.

John Brownlie Henderson's vision over one hundred years ago of a single laboratory servicing the whole of Government continues to be the m+9ost efficient approach. This moved to a new level during the 1990s with the incorporation of the Forensic Pathology, Microbiology and Virology laboratories, and the Health Physics laboratories.

FORENSIC SOIL TESTING

Soil Density gradient created by making series of liquids of different densities in a tube. The dried soil is dropped into the tube and is separated into layers of different density minerals.

THE RECYCLED WATER PROGRAMS

This was a very major event involving analysis of many new compounds. Most sections of the laboratory were involved. During the drought of the early 2000s through to 2011 dam capacity was down to 16 percent. The State built a water grid and desalination plants and recovered water from sewage. Tests were carried out for bacteria, viruses, pesticides, drugs, and other chemicals as well as radiation.

CHAPTER 20: OVER 100 YEARS OF STAFF PHOTOGRAPHS

Laboratory Staff 1908

(Back row) O. H. McNeill, Mathieson, J. McMacGibbon, W. Woodfull, C. Brunnich, J Rait. (Front row) L. A. Meston, F. E. Connah, T. McCall, John Brownlie Henderson, P. W. Jones, H. Macauly-Turner, T. R. Jack.

SENIOR MANAGERS CHEMISTRY

LABORATORY STAFF 2011

Jenny McGowan, Peter Culshaw Richard Mattner, Henry Olszowy, Neville Bailey, Pieter Scheelings, Mark Stephenson, Methsiri Edirisinghe, Gary Golding, Tony Peter, Graham King, Lenore Hadley

LABORATORY STAFF 2011

CHAPTER 21: EMERGENCY CHEMICAL ANALYSIS OF POLITICAL AND MEDIA CONCERN

Apart from the routine work the laboratory also carried out many non-routine investigations.

- Box flat mine (Ipswich) explosion
- Vinyl chloride monomer in food containers
- DDT in human milk
- Cucurbitacin (bitterness) in zucchini
- Crash of King Air aircraft at Charleville
- Brisbane post office explosion
- Doctor misconduct with patients
- Air testing in Emerald where pesticides were thought to have caused a leukaemia cluster
- Cole's extortion
- Queen Street building site accident—cause of lifting sling failure
- Horse racing drug scandal—caffeine in pH dip sticks
- Cyanide in meat pies concern
- Hazardous waste site

- Craigslea State School arsenic contamination
- Mt Isa lead contamination assessment
- Assessment of gasworks sites in South-East Queensland
- Thomas Dixon Centre assessment and clean up
- Cotton store fire at Evans Road
- Esk Main Street truck fire involving pesticides
- Lead in imported toys—child safety
- Glutaraldehyde in Queensland Health endoscopy clinics
- Bray Park and Gold Coast fireworks death
- Heavy metals in children's crayons
- Assessment of environmental problems at Emperor Gold Mine, Fiji
- Tricrecylphosphate contamination in commercial aircraft air supplies
- Ammonia tanker massive 33,000L leak in suburban Brisbane – emergency response and remediation
- Blackhawk helicopter crash investigation
- Assessment of contamination around power poles (Ergon and Energex)

- Conspiracy to murder with cyanide
- Arnott's biscuit extortion
- Heron pharmaceuticals—strychnine extortion
- Geebung gas leak—identification of cause and safety advice
- Air crash near Mount Isa—cause of death
- Leaking shipping container of isocyanates at the Port of Brisbane
- Uptake and depuration of organochlorine pesticides, supplemented into soil ingested by cattle
- Four cities air pollution study
- Pollution survey of Gladstone shale oil plant—pollution minimisation
- Anthrax white powder scares
- Moreton Bay pollution studies
- Bali bombing—disaster victim identification
- Gladstone oil spill—analysis of safety of seafood
- Brisbane water supply dam closures: blue-green algae
- Gold Coast airport hydrocarbon contamination

- Mobility and risk assessment of termiticides in soil
- Asian tsunami—disaster victim identification
- F1-11 fuel tank cleaning investigations
- Narangba pesticide fire study
- Moonie oil pipeline leak at Algester
- Cairns post office contamination closure—identification / advice
- Ship with explosive vapours of carbon disulphide - Port of Brisbane
- Narangba pesticide factory fire at Binary Chemicals—emergency response at scene—identification of hazards—public safety
- Solomon Islands intervention, assisting EHOs
- Condamine Balonne River survey
- Sizzlers extortion—identification of poison
- Poisoning of children's play area
- Recycled water—water grid
- Contamination and risk assessment of arsenic at former arsenic mine sites in Queensland
- Dr Haneef case—analysis of items from investigation
- Chloropropanol in fried foods

- Petrol Tanker overturn-overs and spills Loganholme and Breakfast creek
- Lead contamination at Royal Brisbane Hospital—analysis of lead in air Dust—safety of children
- Asbestos factory fire at Sherwood
- Lead exposure Mt Isa
- Introduction of saliva drug driver testing Southeast Queensland
- Water- Quality Recycled water validation
- Cassava chip recall—analysis and advice
- Narangba industrial estate survey
- Gladstone air shed survey
- Recycled water safety
- Fluoridation of water
- Assisted EHOs with Bali interventions
- Cairns' fish kill
- Lead in children Mt Isa
- Wellers Hill school evacuation
- Mitchelton tear gas evacuation
- Eschem Mine salt contamination of Rockhampton's water supply
- Hendra virus outbreak

- Poisoning of the Tree of Knowledge Barcaldine
- Fluoridation of water supply—analytical control
- Two headed fish case Noosa
- Victorian fires—disaster victim identification
- Major fish kill Maroochydore River
- Moreton Bay oil spill - Seafood clearance
- Amberley RAAF contamination of river with heavy metals
- Contamination issues with the desalination plant—built on contaminated land
- Papua New Guinea air crash
- Investigate a fish kill in a lake at Beachmere
- Anatomical pathology—formaldehyde
- Gladstone air shed study
- Cougha Energy underground coal seam gas
- Brisbane River Oxley Creek flood follow-up
- Sheng Neng coal ship grounding on the Barrier Reef
- Clandestine laboratory house explosion
- Springfield Tomato farm poisoning Bowen
- Coal seam water safety
- Strawberry farm poisoning Beerburrum
- Salt levels in flooded open cut coal mines Central Queensland Post January 2011 flood monitoring

- Gladstone Harbour—fish with lesions
- Interpol cocaine seizure $30m
- Peanuts in toothpaste murder attempt—court case
- Survey of fruit and vegetables for pesticide residue (maximum residue limits) with Brisbane North and Gold Coast public health units.
- Gladstone Harbour: testing of water for 3,4-dichloroaniline (for ALS Environmental) for the impact of coal seam gas expansion
- Queens Park Ipswich—poisoned heritage trees
- SEQ Water analysis of fish flesh, gills, and liver for pesticide residue
- Fuel oil spill in Brisbane River (Hamilton wharf). Analysis of water samples for the extent of spill and containment
- Rapid response to nodularia (algae bloom) incident at Gippsland Lakes, Victoria. Analysis of fish, shellfish, and prawns for algal toxin.
- Several samples from potential/alleged contamination of rainwater tanks from pesticides

- Accidental contamination of a swimming pool with chlorpyrifos (OP insecticide). Samples from the PHU.

- Poisoning of strawberry plants and crop with pesticide—samples from Sunshine Coast PHU
- Survey of organic flours for pesticides for the Public Health Unit
- Papua New Guinea TB drugs strength tests
- Singh triple murder court case
- Wife dissolved in acid murder case

- Multi-agency response Gold Coast fake methylamphetamine case

- National genetically modified food survey

- Dimethyl amylamine food supplement investigation

- A conviction for exposing a child to a clandestine lab in house

- Illicit drugs in sewage

- Canned pineapple extortion attempt

- Antibiotics in imported honey

- Nitrofurans in imported prawn

- Mercury in fish at Hinze Dam

- Needles in strawberries

- Brominated flame retardants

- Nicotine in vaping utensils

TRIACETONE TRIPEROXIDE EXPLOSIVE

A small bottle containing a powder was found in a clandestine methylamphetamine laboratory. When a spot test was performed on a few milligrams to determine if it was an amphetamine, it caught fire and popped. Fourier transform infrared spectroscopy was performed and it

was confirmed to be triacetone triperoxide, a very unstable high explosive. The Chief Inspector of Explosives was called, and he organised for it to be taken out the back of the laboratory in a bucket of sand, placed in a hole, and detonated. This event and others caused a workplace health and safety debate. How does a laboratory which identifies unknown chemicals protect its staff from unknown hazards? The normal process of requiring a material safety data sheet with each chemical does not work with unknowns. The only approach is to assume all unknowns are dangerous and act accordingly.

CHAPTER 22: OVERSEAS USERPAYS ISSUES

In 2011, we received a visit from a Director of an overseas laboratory who gave us the following information about their experience of the implementation of user-pays in a forensic laboratory.

- This Forensic Laboratory had progressed through a series of stages from being part of a government department to becoming a statutory authority.
- The budget was given to the police to spend as they wished. There was a big change in behaviour with a drop-off in document examination.
- The police initially paid for each item after analysis.
- This proved unsatisfactory because there was no way of predicting the demand for services. It was not possible to buy new equipment or recruit additional staff because there was no guarantee of work. Recruitment and training take time and could not be done until the volume was known. This caused significant delays in reporting.
- They told police that they needed to maintain a certain level of work to maintain expertise.
- Police did not wish to pursue a service level agreement.

- In 2006, staff levels were reduced. The backlogs grew and turnaround time extended. They had to outsource work.
- The laboratory was losing money.
- Media attention caused ministerial intervention.
- It was recognised that there was more to forensic science than just dollars and performance levels. This was a turning point.
- The laboratory went from 24 to 47 staff. They set up a satellite DNA laboratory to increase capacity. This proved useful for contingency planning for disasters such as a fire at the laboratory.
- They had to quickly employ and train many people. DNA testing time is now (2012) three to six months.
- They are looking at replacing their laboratory information management system (LIMS), however, every area needs a slightly different set of criteria. Some had need for the calculation, some had a high need for photographs. The competing needs of the different groups complicated the selection.
- They were also tied to the remote central government computer facility via a cable. This significantly slowed up operations. An on-site server may have been more effective.
- They archived all their data every seven days.

- The initial vision was to go paperless and have all instruments connected to the LIMS.
- They initially connected all instruments to the system, and as new instruments were purchased, they were connected as well. This proved to be a source of ongoing cost. They no longer connect the instruments to the system.
- They work on a seven-year instrument replacement cycle. The cost of the interface cards and connections of each new system to the LIMS have proven very expensive.
- Urine testing remained directly interfaced. Every time they purchased a new drugs in urine analyser there is an ongoing cost.
- The system provides management data.
- Certain drug charges are capital offences. This puts a lot more strain on the analyst.
- They issue 600–800 reports per month. They report each item in a case separately rather than one case per report as we do.
- They are looking for a new LIMS system.
- The reports are not fixed by the system. They can be exported into Word so that special edits can be made to the report when required.
- They have an interface with the police with all raw data being loaded by the police before sample submission.
- All the analytical data is sent back to the police computers so they can use it if they wish.
- The representative said they were initially very focused on serving the client at the

expense of training and research. However, they soon realised that the laboratory would suffer long-term with this approach. Now each person has a project or research item in their personal development plan. There is a buffer built into the charge to cover the cost of this research time.

MURDER AND MUD

As part of a police investigation, samples of soil were obtained from the inside of the taxi where a victim was murdered. Examination of the soil revealed the presence of welding spatter, which consisted of tiny, microscopic spherical iron particles. A magnet proved useful in separating the welding spatter from the soil. This information suggested that the suspect worked in a welding works. Neville Bailey received a letter of commendation from the police commissioner for his work on this case.

CASSAVA CHIPS AND CYANOGENS

The presence of high levels of cyanogens in imported cassava chips reported by Japan in 2012 required the rapid development of an analytical method to survey Australian cassava chips. Fortuitously, the laboratory

was able to apply the cassava test kit developed by Howard Bradbury for in-field' testing in Africa and modified it for laboratory use, to participate in the FSANZ national survey. Initial results indicated a high non-compliance with some samples exceeding the temporary FSANZ maximum limit some 10-fold. The issue was resolved by importers securing supplies of sweet cassava with low residual levels of cyano-glycosides

STICKY STRAWBERRY TRAYS

A strawberry packer was having problems with their strawberry punnets. They were sticking together making it difficult for the packers to remove individual empty punnets from the stack. The packer suggested that the manufacturer of the punnets had not used a silicone oil release agent. Surface infrared analysis showed that silicon oil was indeed present on the surface of the punnets. It was observed that the old punnets were of a slightly different shape to the new punnets. The new punnets had a rounded bottom instead of a flat bottom. Apparently, during the forming process the punnets were overheated causing a change in shape. The packer did not believe this result and hired an

independent consultant. When the consultant was told the story by the lab staff, he agreed that this was indeed the case. Easy money for the consultant.

PRACTICAL JOKES

One of the more gregarious staff members wore a blue cardigan to work every day during winter. Every day he took the cardigan off and put on his lab coat. One day when he came to go home, he found the cardigan was missing. Over the next few months photographs of the jumper started to turn up. It was well travelled, having visited Asia and Europe. It was also seen being worn by various members of staff, including the chief executive officer and the Director General of the Department. More concerning, photos showed the jumper being thrown in the garbage hopper or onto a fire. Fortunately, it was returned unharmed!

PAINT CHIP IDENTIFIES HIT-AND-RUN VEHICLE

A chip of paint was found on the clothing of a hit-and-run victim. The microscopic examination showed that it consisted of several layers of

paint and fillers of different thicknesses and colours. When the suspect vehicle was finally located, it had an area of damaged paintwork with some missing paint. The paint on the car immediately adjacent to the missing piece matched the fragment from the clothing in the number, colour and thicknesses of the several layers.

CHAPTER 23: SECTIONAL ROUTINE ACTIVITIES 2012

A wise man makes more opportunities than he finds.

Francis Bacon 1561 – 1624

INORGANIC CHEMISTRY
- Standard water analysis
- Environmental metal pollutants
- Speciation of heavy metal compounds
- Hub for metal determinations by ICP and ICPMS

TOXICOLOGY
- Drugs and poison screens on blood
- Testing related to post-mortem investigations
- Monitoring for the drug courts
- Prisoner urine drug monitoring
- Driver blood alcohols
- Drugs and saliva testing for confirmation of roadside drug positives

FORENSIC CHEMISTRY
- Illicit drug testing
- Clandestine laboratories call outs and remediation

- Physical evidence
- Advice and sampling at clandestine laboratories

FOOD CHEMISTRY
- Regulatory compliance of food including soft drinks, mincemeat, and alcoholic beverages
- Foreign objects in food
- Testing of food colourants, as well as food allergens and fish speciation
- Isotope ratio mass spectrometry hub

INVESTIGATIVE CHEMISTRY
- Volatile organic compounds in the air
- Chemical emergency response
- Pharmaceutical drug analysis including steroids
- Pesticide formulation
- Asbestos identification
- Commercial product analysis
- Toy safety
- Destruction of controlled drugs

CHAPTER 24: PERSONAL CAREER STORIES

RON BILTOFT

Centre Manager Forensic Sciences

1963-1999

Ron joined the GCL in June 1963 as a Chemist Division 2. Up to that point he had been working towards a career as a Secondary Science teacher but realised halfway through his Diploma of Education that he was not suited to that occupation. With a teacher's scholarship and a teacher fellowship granted during his studies, Ron was under contract to work for the education department for six years but was fortunate to be able to transfer his bond (with an extra year to demonstrate good faith) to the Department of Health and Home Affairs.

At that time, the laboratory was located in the Department of Primary Industries building in William Street and had a total of 50 staff. The lab comprised several sections—Foods, Ores, Toxicology, Customs and State Stores, Paints, Waters (which was located in Colchester Street in South Brisbane).

Ron began work in the Ores section, initially on coal analysis. At that time, coal mining was booming, with growing exports, mainly to Japan. The State Government (via the GCL) undertook to do much of the quality control to support the industry. Coal was tested for its specific gravity, calorific value, ash, sulphur and iron content, and other parameters. Other main areas of work were gold and silver assaying of minerals, and soil analyses for metals such as copper, nickel, chromium, tungsten etc for the Geological Survey Office.

A minor part of the work involved testing the stability of gelignite for the Inspector of Explosives. This was done in an old fibro shed on the bank of the Brisbane River, with no protective gear and no fume cupboards, absorption of nitro-glycerine through the skin was inevitable. A morning's testing would produce such violent headaches that staff were unable to work in the afternoon, and Victor Cundith would send them home.

It was the practise to transfer staff around the sections to broaden their experience, and in 1965 Ron was transferred to Toxicology. The work mainly involved analysing samples for alcohol and drugs (mostly barbiturates) to help support the determination of the causes of death in suicides (or the occasional homicide). At that time, testing of drunk drivers wasn't compulsory, so only a couple of hundred tests per year were done. Testing for blood alcohol levels was done by a distillation, oxidation and titration technique which was time-consuming, and throughput was low. In 1968 the Government brought in compulsory testing for all drivers suspected of being under the influence of alcohol. Sample numbers jumped to 1,200 per year, but no extra staff were added. The new legislation required that the analyses be carried out by a State Analyst, in a similar vein to the Food Act. Up to that time, State Analysts had all served at least two years in the Foods section, which Ron had not, but for him to do the work he was deemed capable and hurriedly gazetted as a State Analyst.

Initially, the legislation didn't provide for an analyst's certificate to be prima facie evidence, so State Analysts had to give evidence in court in all 'not guilty' pleas. Ron often would spend up to two days-a-week in court, travelling all over the state, in addition to his normal bench work. Later a gas chromatograph was acquired, and extra staff added, the legislation changed to allow certificates as prima facie evidence, and court appearances dropped off markedly.

After some time, the Government introduced breathalyser machines, which the QPS used to measure blood alcohol levels in designated police stations. Laboratory staff were involved in training the police in their operation, and in supplying certified alcohol solutions for calibrating the machines.

The late 1960s saw an increase in drugs, (mostly cannabis, LSD, and mushrooms, with a lesser number of morphine and heroin cases), and amphetamines became popular in the general population. These cases were covered by the Drug Act, where any contested cases required the State Analyst to give evidence in court. The Drug Act caught up with the Traffic Act and eventually allowed certificates as prima facie evidence. With an increasing workload more staff were appointed, and finally the Forensic Drugs section was split off from Toxicology as a separate section.

When Roy Potter, who had been the Chief Chemist in Customs, State Stores and Paints retired, there was a reorganisation of the sections. State Stores and Paints were separated with John Yule in charge, Don Lecky took his place in charge of Toxicology, John Foreman became Chief Chemist in charge of Customs, but also inherited the work on pesticides and chemical products, pharmaceutical drugs and general health complaints work, which up to then had been done by the Foods section.

At that time transfers between sections were rapid and the affected staff member was not often consulted. One Friday afternoon, Dave Mathers told Ron to report for duty in Customs on Monday. He was delighted with the transfer, as it gave him a whole new field of experience.

After Ron had been there some time, Wally Prentice, the Senior Chemist in charge of the Foods laboratory, died suddenly and the position became vacant. With no experience in foods work, Ron applied, and to his surprise was appointed. Allan Webb, who did have experience, appealed. On the evidence presented,

Allan's appeal was rightly upheld. In the end this was fortuitous for Ron because a few weeks after the appeal John Foreman became Assistant Director. The position of Senior Chemist in Customs became vacant, and Ron was appointed. It was during this time that the lab was fumigated to get rid of the West Indian termites, and staff were accommodated at the Animal Research Institute at Yeerongpilly while the fumigation took place.

Ron was next promoted to Assistant Director, with responsibility for Customs, Foods, Waters, Mining and Pesticides. Ron was in this position when the lab moved from William Street to Coopers Plains. The new laboratory space was a delight once the initial teething problems were sorted out, with much better working conditions. In 1989, under the new Labor Government, a new Director-General of Health was appointed. A major review was conducted and recommended sweeping changes. Many senior positions were put on contract, and the public service became the public sector. Many functions including inspection and regulatory testing of foods, drugs, weights, and measures complaints etc. were wound down and replaced by a system of 'industry self-regulation.'

By the time Trevor Beckmann retired, the Health Department decided that the laboratory profile needed to be lifted, and that staff should become more involved in doing scientific research. The outcome was to appoint a director who was also a university professor—that would have joint responsibilities, and both organisations would benefit from mutual interaction. Des Connell was appointed. This philosophy continued with the appointment of the next Director, Michael Moore, another university professor.

The Department appointed a CEO as Director of the larger organisation. He was a skilled political operator and was able to achieve quite a bit for the lab.

In the early 1990s, the laboratory was becoming more involved in chemical contamination issues including a major arsenic contamination problem at a local school. To gain more expertise, Ron enrolled in a Master of Environmental Engineering at Griffith University specialising in Waste Management.

Around 1992 the new mortuary had been built on the Coopers Plains site, and after some time the Laboratory of Microbiology and Pathology section moved out of their premises in George Street (Brisbane). Microbiology was allocated space in the main building, and Forensic Biology was housed in the mortuary building.

When the mortuary first moved out to Coopers Plains there were protests by some of the local residents, with coverage by the local media. There were references to mortuary activities and concerns about infection problems. It was agreed that the hearses would not travel down Middle Street but enter and exit the precinct via Kessels Road. A community liaison officer was appointed to help families with information and assistance to deal with the stress and trauma when autopsies had to be performed.

Time was spent working out the costs of analyses completed in the lab when introducing a fee- for-service. A large compilation of codes was developed and computerised, and each month clients were advised of the value of work done for them. In some cases, clients were charged real money, and the system was effective in reducing workloads, but for many clients, especially in forensics, where analyses were prescribed by law, the charges were only notional. The clients had little control over what had to be done, and little change occurred.

One day, after over a century in existence without ceremony, the GCL ceased to exist. The next phase of reorganisation was to combine the GCL and the LMP into one organisation, QHSS. Henceforth it was verboten to mention the GCL by name. It was often secretly referred to as the 'three letter acronym' by staff. There was no celebration of the change.

The new CEO amalgamated all the hospital pathology laboratories, which to this stage had been under the control of the pathologists in each hospital, into the single organisation called Queensland Health Pathology and Scientific Services. Thankfully, they were never closely allied with us. However, it did bring enormous benefits in standardising testing and reporting and saved large sums by centralising purchasing of supplies and equipment.

The new combined labs were then split into three Centres, each under a manager. Although Ron had just completed his Master degree in Environmental Engineering, he became Acting Manager of the Forensic Science Centre, in charge of Pathology, Forensic Biology, Toxicology and Forensic Drugs. Pathology and Forensic Biology had a very different culture to Toxicology and Forensic Drugs, and it took a lot of effort to form a cohesive group, especially as they were split between two buildings.

The most pressing problem facing the Forensic section was severe under-resourcing. Ron prepared a lengthy report for the CEO detailing the problem, and an assessment of what resources were required to meet the section's responsibilities. The CEO's management style was to vigorously challenge the recommendation of a subordinate. He accused Ron of trying to "snow" him, but after some persuasion he submitted his report to the Victorian Forensic Sciences for comment. They told him Ron's figures were reasonable, and that they couldn't believe we were doing so much with so little. It must be said that many staff were working large amounts of unpaid overtime to achieve this feat.

The Forensic sections also suffered a fundamental issue in being part of the Health Department. It was not core business of the department. While most of their work was generated by the Police Service, our results were required for the Justice Department. The laboratory had no control over the workload and were the meat in the sandwich.

The Health Department was always under financial pressure and wasn't keen to put extra funds into the lab. The CEO did manage to get the Director General of the time to come to have a look at our lab and work. His reaction was: "That's all very impressive, but what's it got to do with health?" Over the years, Cabinet had considered whether Forensic should be transferred under the control of the Police Service, but the legal profession was strongly in favour of having it remain independent of the Police, and it was left in Health.

In an effort to achieve better understanding and co-operation between all the parties involved in the Justice system, and with the strong support of the CEO, Ron managed to get an Inter-Departmental Committee for Forensic Services formed with Police, Prosecutors, Coroners and Health representatives. This brought some good results, although they were slow in coming until he built up some level of trust. At one point the CEO offered to transfer Forensic to both Police and Justice along with the level of funding provided by Health at that time. Both quickly declined the offer.

The CEO also decided that all computer services would be controlled by his office, and the lab lost control of our own computer staff, who had all worked in laboratories as scientists and understood very well what we required. A new laboratory management system was developed which suited the pathology laboratories but was completely inadequate for most of the chemical laboratories, particularly Forensic. The

CEO would not be persuaded and accused all the chemists of being recalcitrant and simply resistant to change. He once said to Gary Golding "you might as well face it you are going to get AUSLAB" even though the author had previously volunteered to be the first section to trial AUSLAB so that it could be better tailored to the chemistry lab's needs. A communication failure.

The central computer staff did not understand the lab's requirements. On one occasion at a meeting of all the QHSS managers, Ron publicly criticised the new computer system, and the CEO was reported to say, "That's the end of him." However, despite his misgivings about Ron's approach to several issues, the CEO arranged for him to be awarded a Queensland Health Australia Day medal "For Achievement" in 1997.

The Department had determined that the lab should get Quality Management Accreditation in Forensic Science, with a time allocation of one year. It was a formidable task, with considerable resistance from some staff, who were already totally overloaded with work, but with a lot of cajoling and support they managed to achieve it. Once it was in place, the benefits were clear, and it was well-accepted by the staff. Peer review by staff of all results was widely supported as errors were found and corrected by this process.

In 1994, it was decided that the lab would provide a chemical advisory service to Emergency Services to assist the Fire Services and Police in dealing with fires and other events where contamination was a potential problem. Ron was one of a few volunteers. They received training from the Fire Services in wearing breathing apparatus, moon suits etc, and were provided with a van equipped with a range of monitoring equipment. A member of the team was rostered on call 24/7. He only did this for a couple of years – as his work with Forensic became too onerous and he had to give it up, but it was an interesting time.

In 1999 it was announced that all the Manager positions were finally to be advertised and filled. Having acted in the position for 5 years, by that time he had had enough, and chose not to apply, but to retire, which he did in July. He had worked in the lab for 36 years, most of which he said he enjoyed immensely.

KINGSTON CONTAMINATED SITE - MEMORIES BY RON BILTOFT AND GARY GOLDING

Important things can start in small ways. Gary recalls: a phone call received at the laboratory from a lady from Kingston who said there was black material bubbling out of the ground in her garden. When she said the material was carried indoors on shoes and that it was dissolving her carpet, things seemed a bit more serious. Gary asked her what her carpet was made of, and she said Nylon. He knew that acid dissolved nylon. Tests on the sample showed it was a mixture of a bituminous material and sulphuric acid.

Ron recalls: When the saga of the ground contamination at a site in Kingston began, he thought he was still Senior Chemist in charge of Customs. The contamination arose from drums of chemical waste including used engine oil, and other waste products, dumped into an old gold mine. Over the years the drums rusted, and the waste spilled out and came to the surface in what was by then a housing development. The lab had done huge amounts of testing to determine the nature and extent of the contamination. It was highly political, with several residents claiming it was causing health problems, and a lot of media involvement. A remediation plan had been developed, but before it was put into action the Goss Labor government was elected. The new government immediately declared that all the previous work was suspect, and testing started again. Apart from some testing going to Greg Miller's lab, most of the work was redone by the GCL, with pretty much the same results. Overall, it cost the Government millions of dollars. In the end most of it was covered with bitumen and became a car park.

 The National Measurement Act

One interesting situation arose with the legality of a blood alcohol measurement. The lab had for some years been using certified weights and a certified reference ethanol to standardise our equipment used for blood alcohol analysis. The results were quite accurate, but unbeknown to us, the National Measurement Act, which the State Government had accepted as Queensland law, specified that *all legal analyses had to be done using reference standards certified by the National Measurement Laboratory in Canberra. This meant that many of our analyses, including blood alcohols, were technically illegal. This fact came to light in a court case, causing some consternation. Ron had to advise the Minister and seek to get the law changed retrospectively to avoid hundreds of convictions being overturned. While sitting in the Ministers builder waiting room to see him, it suddenly hit Ron Biltoft that the issue was much wider than our analyses – it also affected any measurement made under any regulation or contract – fuel volumes, food weights, coal contracts, water meters, speeding fines etc. etc. The matter was fixed almost overnight.*

GEOFF RYNJA FRACI

Assistant Director 1967-2005

In 1965, Geoff Rynja was awarded a scholarship to study the new degree level chemistry program full-time at the then Queensland Institute of Technology (QIT). He became one of the program's first graduates. After graduation, he was bonded to the laboratory and commenced work full-time in December 1967 in the Toxicology laboratory. There, he analysed toxic heavy metals in biological fluids and was introduced to post- mortem toxicology.

In mid-1969, he was transferred to the Mining and Industrial Hygiene laboratory. He participated in the analysis of a very large number of geochemical, coal, limestone, clay, and oil shale samples from various government resource mapping programmes plus the weekly samples from mineral prospectors. Staff numbers swelled to 23 members, who used conventional methods and wet chemistry techniques. The workload led to the eventual purchase of instruments that allowed the simultaneous analysis of many elements.

During the period 1968–1972, he studied part-time at UQ and obtained a Bachelor of Economics degree.

In 1971, Geoff began an involvement with industrial hygiene. He was selected to be part of the Queensland Health team to carry out dust and other monitoring in the Mt Isa underground mine, and metals monitoring in the smelter areas. As a follow-on, he was asked to lead teams to take samples from every underground and open cut coal mine in Queensland during 1972–73. The program included roadway dust, respirable dust, and seam gas sampling. As the lab had no experience in underground coalmine work, he took a sampling team underground at Box Flat, Ipswich, to assist with project design. The team was shocked and saddened three days later when the mine exploded with significant loss of life. It was suggested that the mine was already heating at the time of the visit.

The results of the respirable dust field work from these projects were used to support laboratory- based experiments which examined the relationship between respirable mass and respirable surface area. The results were published in an international peer-reviewed journal.

In 1975, the director sent Geoff Rynja for a 12-month tour of duty around the laboratory to broaden his skills base. In that period, he worked in the pesticides and spectroscopy laboratories before returning to the mining and industrial hygiene laboratory. In June 1977, Geoff applied for promotion to Senior Chemist in accordance with the award requirements and was duly appointed as such.

In the mid-1970s, Geoff and other chemists became increasingly involved with oil spill analysis. This arose from a change in state legislation, substantially increased fines, and the appointment of oil spill inspectors. In 1978, Geoff was made Senior Chemist of the Government Contracts laboratory and was instructed to take the oil spill analysis equipment with him. Here, in addition to contracts work, he did some physical evidence work involving petroleum products for the Forensic Chemistry laboratory.

In 1980, he was appointed to the position of Supervising Chemist of the Forensic Chemistry and

Toxicology laboratory. At the time there was a centralised record management system for exhibits. Inspired by issues arising from the Fitzgerald Enquiry, he set about consolidating exhibit management and storage processes. In the mid-1980s, well before Forensic laboratory accreditation began, he initiated work to have all routine analytical methods standardised and validated.

Geoff initiated the formation of the Queensland Branch of the Australian and New Zealand Forensic Science Society (ANZFSS) and was its foundation president in 1982. During the period 1982–1993, he was Honorary Secretary for the RACI Queensland Branch and was elected as a fellow in 1987.

With the advent of the 'user-pays' environment, Geoff was part of Don Lecky's team in 1981 that reviewed and developed a costing model for analyses and services provided by the laboratory. In 1982, he joined Des Connell's team to develop a corporate plan for GCL.

In 1988, he became the Acting Branch Assistant Director Natural Resources and Forensic Services Branch and was in this position when GCL relocated to Coopers Plains.

During 1990 he was transferred to the Environmental Waters and Wastes laboratory where, ultimately, he became the Supervising Chemist. In the ensuing four years, he spent a total of 12 months as an Acting Assistant Director, Natural Resources and Forensic Sciences and 10 months as Acting Assistant Director,

Laboratory and Administrative Services. This latter acting appointment occurred when the incumbent was sent by the Chief Health Officer on secondment to LMP.

In 1995, he was part of a team that restructured the Chemistry laboratory. The restructuring took place in 1997 and saw sections amalgamated, and the roles of assistant director disappear along with several supervising scientist's positions. Geoff's role changed, and he looked after various high-end projects as well as elements of marketing and business development.

From 1995–97, he was Acting Assistant Director, Environmental Protection. During this latter period, he was involved in international aid and development projects, one in Indonesia and one in Papua New Guinea. He also worked on his PhD 'An investigation of pathways to continuous environmental improvement in analytical laboratories'. He used laboratory data in the case study. At the same time, he became accredited as an Environmental Auditor (ISO 14000).

Michael Moore gave him a science-based project relating to pharmaceuticals in sewage effluents. Geoff was given some peer reviewed papers and asked to apply them concerning the use of prescription drugs in Australia. He compiled a comprehensive report which was the basis for a major new research direction for EnTox. This work was extended to include illicit drug use in the community and prisons and continues to this day.

His career demonstrated many innovations. He conceptualised the need for the laboratory to develop skills in forensic transferred fibre investigations. He encouraged Gary Golding to do his master's degree in this field. He also conceptualised the need for a laboratory that could test for volatile organic materials in the air. Once again, he encouraged Gary Golding to implement this initiative. Following the acquisition of the equipment Elizabeth Christensen developed the capability. The laboratory had considerable capability determining organic contaminants in food and water, but with the departure of the occupational hygiene laboratories to another government department, it lost its air analysis capability.

Geoff Rynja was involved with interesting work in the United Kingdom in 1973. He had a Letter of Introduction from the Queensland Chief Inspector of Coals for Her Majesty's Chief Inspector of Coal Mines in London who arranged for him to visit their coal mine dust explosion facility in Buxton. They set off a coal mine dust explosion for him to observe! He also visited the Safety in Mines Research Establishment (SMRE) in Sheffield. The lab used SMRE horizontal precipitators to measure the respirable mass of the coal dust. Geoff retired in 2005.

A SMASHING TIME

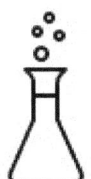

The new spectroscopic laboratory contained the only modern glass door in the building. The renovation had been championed by Keith Deasy. One day in passing the door an unnamed senior lab tech bounced an orange off the door and to his dismay it shattered. The entry never looked the same again.

THE EMPTY BIRO

At some stages, the budget was so tight that staff had to return an empty biro before they were issued with a new one.

BURNED LIPS AND DISSOLVING CUPS

A notable case involved burnt lips on aged residents and dissolving polystyrene cups. This was due to fruit juice containing excessive limonene, which was added during the juice manufacture. Limonene is extracted from citrus skins and when concentrated is also a solvent and cleaning agent.

GRAHAM CRAVEN

Manager Business Support Services 1968-2007

Graham Craven graduated from QIT and commenced work at the laboratory in 1968. He recalls that his interview consisted of one question: 'When can you start?' This was during the time of the mining boom when some experienced chemists left for better wages in the mining industry. He spent a good portion of his early career carrying out food analysis and analysing legal samples for prosecution. As his career progressed, he became the Assistant Director overseeing the work of three different groups, Foods, Pesticides and Manufactured Products. With restructuring in 1995, he took on the role of management of Business Support Services. This covered a wide variety of tasks including building maintenance, budget, administration, library, human resources and workplace health and safety. He was responsible for the day-to- day running of the laboratory during Michael Moore's time as Director. This was a similar role to the one carried out by Don Lecky during Des Connell's term as Director.

Graham was instrumental in initiating, facilitating, and supporting several strategic initiatives. These included the co-location of the LMP Microbiology laboratory on to the site and the co- location of the

Health Physics group. Later in his career Graham became involved in the negotiations to co-locate the DPI laboratories and the CSIRO food laboratories onto the site in newly constructed laboratory buildings. The security levels in the building were gradually enhanced. Security incidents at forensic facilities around the world heightened our awareness of potential dangers. External security was enhanced including plastic films over windowpanes, security gates and obstacles, (large rocks) around the boundary to inhibit vehicle access. There were also internal security enhancements.

During the aftermath of the Black Hawk helicopter crash and fire, many bodies were received at the mortuary. Graham saw the need and arranged for psychological counselling for the mortuary staff. Working with Dan McKeown he overhauled the electrical services. Staged start-up of fume cupboards and air-conditioning reduced peak load upon which the electricity bill was based. The emergency generator was wired to the grid and was run periodically by the electrical generation authority, to ensure it functioned when needed. An uninterrupted power supply was installed to provide power long enough to maintain instruments until the emergency generator came online. Lights were equipped with movement sensors and turned off if no movement was detected. They often turned off during meetings….. followed by much arm waving until they came on again. Graham Craven retired in 2007.

GOVERNMENT PAINT COMMITTEE

Laboratory staff members were also auditors under the Government Paint Committee (GPC). This was a federal government committee with state sub-committees. The staff periodically inspected factory operations to ensure that good manufacturing practise was being followed. It was not unusual to find problems in manufacturing ranging from failure to provide static discharge leads on drums of flammable liquid, to unsealed containers.

The committee was around for about 50 years. This simplified contractual arrangements but relied on the honesty of the painter to use the correct paint. Government inspectors took samples of paint which were forwarded to the laboratory for testing for density, non-volatiles, ash, and lead content. These results were compared with the paint specifications supplied by the manufacturer. There was a small percentage of cases where the paint had been diluted with solvent. Oil-based paints were preferable because they could be scraped off easily when the time came to remove them. Acrylic paints, which lasted longer, were very difficult to remove. Eventually the department decided to go over to acrylic paints and testing procedures were no longer valid. One day about five o'clock in the afternoon, Premier Wayne Goss announced on radio that he had closed many unnecessary government committees, specifically mentioning in a derogatory tone "the Government Paint Committee." This created an issue because the government paint committee (GPC) paints were specified on contracts. The committee was subsequently reformed and called the Surface Coatings Committee. It continued to meet about twice per year saving the Government millions in faulty paint jobs.

GARY GOLDING OAM FRACI

Managing Scientist Chemical Analysis 1969-2013

Gary grew up in the small town of Urunga, New South Wales. He was inspired to a life of science at an early age, somewhat influenced by a news reel movie featuring the rocket scientists at Woomera. Around the age of nine, his father bought him a chemistry set for Christmas, and he spent many hours in the garage carrying out the various experiments. Little did he know that at a later stage in life he would be a member of the Standards Australia Committee on Toy Safety which introduced the Australian Standard for chemistry sets (AS 8124.4 - 2003).

One day, his father mentioned that industrial chemistry was a very well-paid profession. Never let it be said that money has nothing to do with career decisions in science. These events established his course in life. His family moved to Brisbane in 1963 to provide the children with better educational opportunities. He attended Sandgate State High School, which was fortunate to have received a new laboratory building as part of the Federal Government initiative to stimulate science and was one of the trial centres for the Chem Study programme. He did very well in chemistry and physics in his senior examination and received one of only two scholarships from the Queensland Government to study chemistry at QIT.

A condition of the scholarship was a four-year bond to work at the GCL. His career direction was set. The scholarship included working at GCL during the Christmas holidays in 1969 and 1970. This work benefited the laboratory as well because this was a period of high recreation leave levels among the full-time staff.

He commenced full-time work in December 1971, initially in the ores and industrial hygiene laboratory.

In 1975, six chemists were selected to spend a year working in various parts of the laboratory. These chemists were seen as future managers. Gary was one of these chemists. During this time, the staff were expected to gain a good working knowledge of the work of other departments within the laboratory.

He spent the next 42 years working for the organisation through several name changes including QHSS and QHFSS. During this time, he rose from his initial position as a Laboratory Assistant to become Manager of the chemistry laboratories. The ever-changing nature of the work at the laboratory meant that he worked in several fields including geochemistry, water analysis, occupational hygiene, mine safety, environmental residues, food analysis, forensic toxicology, forensic drug analysis, government contract assessment and forensic physical evidence. He was appointed a State Analyst for Queensland and had considerable experience as an expert witness.

One memorable job in his career was the survey of Queensland coal mines for dust explosion hazards. One mine, Box Flat, exploded the day before a scheduled visit, sadly killing many miners.

He served on many state government committees including those involved in the assessment of contracts for cleaning chemicals and textiles, drugs and poisons regulation, counterterrorism, and chemical warfare agent analysis.

Gary's early career involved a large amount of classical wet chemical analysis. With the introduction of instrumental techniques following graduation, he returned to QUT and completed a Graduate Diploma in Chemical Analysis to upgrade and update his chemistry skills. During his mid-career, he completed a Master of Science in Analytical Chemistry (QUT) specialising in forensic trace fibre analysis. As he moved into a managerial role as the supervisor of the Investigative Chemistry section, he completed an MBA in Technology Management. He found this very useful when dealing with external consultants brought into the laboratory to 'improve' operations.

During his career, he was responsible for the introduction of new technologies to the laboratory including FTIR, volatile organic compound analysis in air, forensic fibre analysis and isotope ratio mass spectrometry.

His generalist chemistry skills acquired during this period resulted in him becoming a resource of advice for younger chemists. He was a foundation member of a team of volunteer on-call chemists whose role was to attend chemical emergencies in Queensland to advise the Queensland Fire and Rescue Service. He always felt that this was a worthwhile and fulfilling role. He received an Australia Day medallion as recognition for this activity, and later another

Australia Day award in recognition for services to chemistry.

He served as president of the Queensland branch of RACI for two years (2007-08) and was on the Queensland RACI Analytical and Environmental Group for around 20 years with several years as Chair and Secretary. He also served on the RACI National membership committee. He was elected a Fellow of the RACI. In 2020 he was awarded a RACI Distinguished Fellowship. In 2024 he received an Order of Australia Medal for services to Science – Chemistry.

Following his retirement, which was preceded by a diagnosis of Parkinson's disease. He took part in many Parkinson's research projects at various universities in the Brisbane area. He often joked that he had started out as a chemist and ended up as a sample. He published a peer reviewed paper in this field. Gary Golding retired in 2013 after 41 years of service to the laboratory. Years later he found out that some senior chemistry staff referred to his time as Manager as 'The Golden Years.' He really appreciated their comment.

INGE SCOTT {NEE SCHOTT}

Senior Chemist 1982-2012

Inge Scott was a Senior Chemist and State Analyst with extensive experience in analytical chemical testing in food chemistry, forensic toxicology, and occupational hygiene sections of FSS. Inge started employment at the GCL in August 1982 and continued until December 2012. After 2012 Inge retrained and became a science and mathematics teacher for secondary and tertiary students. Before starting at GCL, Inge was dux at high school (Brisbane) with chemistry, biology, physics, mathematics, and art being her favorites subjects. She has a Bachelor of Science with double majors in both Chemistry and Biochemistry from UQ. Inge won a National Heart Foundation Vacation Research Scholarship for Biochemistry at UQ. She also won the Postgraduate Vacation Research Scholarship from ANU for Organic Chemistry.

At ANU Inge undertook research specialising in antibiotic synthesis for Dr Rod Rickards and Dr Anna Becker. Dr Rod Rickards worked with the famous Australian Organic Chemist, Arthur Birch, (the Birch Reduction enables the modification of steroids) for many years. The following year, Inge won first prize in her Honours year for the Bachelor of Science with First Class Honours in Chemistry. After completing these qualifications, Inge worked as a postgraduate research assistant in Organic Chemistry at UQ for Dr William Kitching.

Inge Scott worked at the Queensland Health laboratories for 30 years. She initially worked in the Forensic Toxicology section. The laboratory tested whole stomachs for drugs and poisons and macerated the livers for drug testing. Some of the blood samples taken by police from drink drivers had levels of alcohol above 0.4-0.45 per cent.

While working in the Forensic Toxicology section in the early 1980s, Inge regularly received post-mortem samples from Papua New Guinea (PNG) on Friday afternoons, as at the time PNG did not have adequate forensic laboratory capabilities. Queensland provided this service due to historical commitments. It was with excitement and trepidation that we opened these containers. Unfortunately, many of these samples were not refrigerated enroute. On occasions the lid flew up several meters when opened. Luckily, the lab had excellent fume extraction cupboards in the newly refurbished toxicology section facing William Street. The laboratory was able to separate the early gas chromatography peaks due to putrefaction compounds from most of the drugs in our detection procedure.

A larger occupational hygiene section was set up in the mid-1980s to monitor and protect workers and the public from exposure to health risks like dust, silica, asbestos, particulates, volatile chemicals, and heavy metals. Monitoring was manual, personal air monitors were fitted to workplace volunteers for several hours to test air quality. Later, mobile infrared detectors were carted to the testing site in the laboratory station wagon until purpose-built vans were commissioned several years later. Larger jobs included mine dust monitoring, automobile brake repair business asbestos monitoring, hospital volatile organic compound (VOC) monitoring and extensive monitoring of a foundry at Runcorn. This survey tested VOCs emitted from the casting process and particulates. Generally, the conditions were within legislated levels. Inge worked in the foods section at the William Street laboratory, this section was located at the south-west facing wing. It was very hot in summer as there was no air-conditioning in the main laboratory. Staff wore long lab coats, safety glasses and gloves.

The laboratory contained several furnaces, drying ovens, Bunsen burners, hot plates and the Kjeldahl room which had boiling sulphuric acid and sodium hydroxide. At the Coopers Plains laboratory, the staff had the pleasure of much better working conditions provided by the purpose-built laboratory. Inge conducted food compliance analysis, food labelling surveys, food complaints (ranging from frogs and insect parts in salad mixes to vegan foods containing pork rind), allergen detection and contributed to several APFAN training conferences.

Many foreign objects allegedly found in food were submitted for analysis including pieces of metal containing mercury, silver, and tin. These foreign objects were consistent with broken pieces of dental amalgam. Food complaint enquiries involved liaison with several other chemistry, histology, and biology sections within FSS to provide an accurate report and solutions to problems experienced with our food supply.

The new high-resolution stereomicroscope with a digital camera and computer accessories played an essential role in the identification and recording of foreign matter. Inge found the work in the food complaints section fascinating, involving sophisticated problem-solving techniques and very rewarding.

Consultation with the environmental health officers aided the investigation processes.

DANIEL JOHN WRUCK

Senior Laboratory Technician 1971-2008

(In Memory)

By Mary Hodge (Obituary originally published in Chemistry in Australia, Vol. 76, No. 3, Apr 2009: 23). Dan Wruck commenced work with Queensland Health in January 1971 as a Cadet Technician at the GCL in William Street, Brisbane while at the same time studying at night, for a Certificate in Chemistry at QUT (then known as the Queensland Institute of Technology). He gained his Certificate in Chemistry in 1973. For the next 10 years, Dan gained experience in various sections of the laboratory performing chemical testing on coal, water, trace metal analysis in water and pesticide analysis.

. By 1985 Dan had gained an in-depth knowledge of High-Performance Liquid Chromatography (HPLC). He adapted this then new technique to the testing of sugars in food, drugs in clinical specimens and pesticides and thus ensuring the successful application of this technology to routine laboratory processes. Dan not only trained government staff in this technique but provided workshops to visiting scientists from Asian countries as part of the Asian Pacific Food Analyst Network (APFAN). In 1992, Dan provided lectures and training on HPLC to Indonesian scientists at a National Indonesian Seminar in Bandung.

In 1988, Dan commenced work in the Environmental Nutrient section of the laboratory, testing low level nutrients in environmental waters. It was in this discipline of chemistry, that Dan became most passionate, becoming a leading expert in this field of science both in Australia and overseas. Dan quickly saw the need for Australian laboratories to provide high quality testing in this area of science and was the coordinator of, and driving force behind, the National Low-level Nutrient Collaborative Trials which provided Australian and international laboratories with a program for inter-laboratory comparison of testing results in environmental nutrients. More recently Dan provided a similar program to scientists in China and Korea. Dan was a member of the Standards Australia and Joint Standards Australia / Standards New Zealand Technical Committee. He was a member of the Royal Australian Chemical Institute and an assessor for the National Accreditation Testing Authority (NATA). Dan provided technical advice and represented Queensland Health in the Southeast Queensland's Healthy Waterways Ecosystem Health Monitoring program. Dan continued his involvement in training, providing training workshops in nutrient analysis both in Australia and overseas including Vietnam, Korea, and Thailand.

In collaboration with the Queensland Department of National Resources and Water and the Environmental Protection Agency, Dan provided workshops to community groups throughout Queensland demonstrating best practise in terms of collecting, preserving, and testing of water samples.

Dan seemed indestructible and full of energy. Only days before his death, Dan was accepted into a PhD program at Griffith University. Dan provided leadership, enthusiasm and common sense to environmental nutrient testing and metrology in

Australia. Dan Wruck died suddenly after a tragic accident at his Mt Mee farm on Wednesday, 27 September 2008. Dan had worked for Queensland Health in its Brisbane laboratories for 38 years. His sudden death came as a shock not only to work colleagues at Queensland Health but to scientists and collaborators throughout Australia and Asia.

 STORED CHEMICALS

In a laboratory involved in a wide variety of one-off sample types, there is a tendency to build up a large inventory of chemicals. Chemists, being chemists do not like to throw these out. If needed, they are required at short notice and the client cannot wait weeks for the new reagents/standards to arrive. One morning, one of the staff was having morning tea and noticed that his lips were swelling. This was soon after he went to the drawers containing drug standards. A quick look in the drawer showed a small bottle, with a ground glass lid, which had spilled in the cupboard. The label was aconitine. His symptoms were typical of aconitine toxicity. Having a wide variety of chemicals on hand can be lifesaving. Gary Golding recalls getting a call, one weekend, from a Mater hospital doctor. They were looking for some strychnine to treat a baby with a rare problem that required a small dose of strychnine. He was able to supply some after a bit of searching.

John Vivian Foreman
Deputy Director 1946-1986

(In memory)

By Gary Golding (Obituary) originally published in Chemistry in Australia, v.73, no.10, Nov 2006, p. 27.

John Foreman or Jack as friends and family knew him, was born in Maryborough on 7 November 1927. John spent his formative years attending Maryborough Central School then Maryborough High School until 1942. In 1943 at the age of 15, John was sent to Brisbane as a Payroll Clerk in the Department of Health and Home Affairs. He was to work in the Queensland Health department for the next 45 years. John was soon elevated to the heady position of Clerk in the Queensland GCL and was later invited to become a Cadet Chemist. John attended UQ from 1945 to 1950 where he studied science at night, while working full-time. Upon his graduation, he became a Government Analyst, a position subsequently known as Chemist. In 1948 Jack met his lifelong partner and wife Dorothy on a week-long bus trip to Armidale. He spent his career applying his analytical chemistry skills to the solutions of the many problems facing the

Queensland Government and the Australian Customs Service. Those he managed were constantly in awe of his broad knowledge and practical approach to chemistry. Towards the end of his career, John took on the role of Deputy Director of the Queensland GCL. He retired in 1988, and this gave him more time for other activities such as involvement in the Queensland Public Service Retired Officers' Association and in Freemasonry. In retirement, John liked to brew beer, follow the stock market, and read the paper.

In our lives, we look for guidance from those we know who do things better. Those who learnt their practical chemistry from him recognise his outstanding leadership, management, and chemistry skills. He was a decent and fair man who, through thick and thin, kept in true perspective the relative importance of things. John will be sadly missed by his family and his chemistry colleagues.

John joined the RACI in 1960 and remained a member for 46 years until his death on 1 March 2006. He thus maintained an interest in chemistry to the end. Thank you for everything you were and everything you did.

IDENTIFYING STOLEN TIN

A significant quantity of tin ore went missing from a dredge in North Queensland. It was suspected that the tin had been received by a tin dealer who also bought tin ore from many tin scratchers working in the area. Samples were taken from a number of kegs of ore on the premises and submitted for comparison with ore from the dredge. Analysis of the particle size profiles, and the colour and types of minerals present were inconclusive. However, passing a magnet over all the samples yielded particles of steel, which on analysis proved to have an identical composition to the dredge bucket steel.

HIDDEN BENEFITS

John Foreman's home brew was said to taste just like the local brew 4X. Coincidently the lab had received a sample of the brewery's hops extract. Several home brewers had their eyes on the residual sample and would have been happy to "dispose" of it following analysis. One day it disappeared and was not seen again.

SYNTHETIC HASHISH

The laboratory decided to carry out a project to look at the trace metal content of hashish in the hope of becoming able to profile the origin of samples. The initial step in the operation involved ashing the hashish in a furnace. The exhaust system covering the furnace was not particularly effective and a strong smell permeated the entire laboratory. Most of the samples were about 21% ash – one was about 33% ash. An infrared analysis of the ash showed that this sample contained talc. It was probably a mixture of hash oil and talcum powder. This would make the product more dangerous as talc is known to be a lung irritant and can cause the disease talcosis which is like asbestosis. This finding was published in an international journal.

DR MANNIE RATHUS

Although not a staff member of the GCL, during the 1970s the laboratory worked in close cooperation with Dr Mannie Rathus who headed a small occupational safety unit with Dr Chalk—son of the Deputy Premier. He was a South African military doctor and an ex South African title holding boxer. Being a doctor, he had to stay behind the lines during the Second World War. Scuttlebutt has it that he stowed away on a bomber as a rear gunner and shot down an enemy fighter.

He was responsible for workplace issues and, in collaboration with the laboratory, periodically investigated new aspects of worker exposure to dust and volatile chemicals. In 2018, a report he wrote on black lung in miners in the 1980s became topical although the report appears to have been ignored at the time. He was an outgoing character who did not think it abnormal to yell out to a shy young chemist, across busy lunch time Queen Street (before the mall), asking about a particular set of results. Without his enthusiasm for occupational safety the laboratories involvement in the field may have been substantially less.

He instituted a survey of blood lead levels in children, which was published. He took an active role in the fumigation of inner-city buildings monitoring methyl bromide in the blood of workers by XRF (for bromine levels). Some years later, one of his own blood samples was found in a freezer marked by one of the staff: keep for cloning.' He was a unique and highly respected person.

TIME MANAGEMENT

Gary Golding attended a Time management course. It commenced at 9am. The group was asked at 9:15am what their biggest time management issue was, Gary replied, constant interruptions,' almost immediately there was a knock on the door and a person came in and said there was a telephone call saying he was required in court immediately. Later he was enrolled in another time management course and a more important issue arose, so he sent one of his staff. Delegation is a good time management tool.

RADIOACTIVE SAND AS TOP-DRESSING

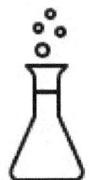

In the late 1970s a company sold mineral sand mining waste to house holders to top-dress their lawns. The waste contained the radioactive mineral monazite. After the media exposed the story, there was public outrage. The government tried to reduce the outrage by claiming, at a public meeting, that the increase in exposure in the houses was not detectable. Unfortunately for the speaker, one of our analysts was in the audience and he had made some measurements that demonstrated that this was not true. He raised this at the meeting and was questioned about where he got the equipment. He had borrowed it from the lab. Instead of accepting that he had shown initiative to protecting himself and his family through his readings, the laboratory was asked to explain why we allowed equipment to be borrowed by staff. Eventually all the contaminated lawns had to be dug up and the topsoil replaced.

COAL WASHING PURGATORY

If for some reason you upset management in the mid-1970s, you were sent to the coal washing area. 200L drums were filled with zinc chloride solution of various specific gravities. The coal was put into the drums and the fraction that floated measured. The solution was heated on a gas ring burner. Zinc chloride was deliquescent and corrosive. The washed coal was dried in the sun on sheets of galvanised iron. Not a pleasant job.

HENRY OLSZOWY

Chief Chemist Inorganic Chemistry 1968-2012

Henry commenced work at the laboratory as a Laboratory Technician following completion of the Certificate in Chemistry course at QIT in 1968. He then went on to become one of the more highly qualified members of the staff. He completed a PhD Degree in Chemistry from UQ in 1984. Over 20 scientific publications in prestigious journals resulted from these studies. Henry also completed a Master of Engineering Science in Waste Management at Griffith University in 1995.

Henry accumulated over 40 years' experience in analytical chemistry in a diverse range of disciplines including occupational hygiene, biological sciences, food chemistry, water chemistry, forensic investigations, manufactured products, environmental toxicology, and general classical chemistry.

His work had a public health focus involving waters, environmental and health programs. Henry was also a State Analyst for Queensland Health and had experience as an expert witness for the State.

Henry has been a member of many committees including the Chemical Pathology Discipline Working Party.

More recently, he became a committee member on the Australian Standards Committee (CS18) on Toy safety,

Chairperson of a Standards Australia Committee on Spectroscopic Analysis CH/16 (revision of AS 2134, AS 3641, developing AS4873.1 and AS 4873.2),

a committee member for the Standards Australia Major and Minor Components in Coal Ash by XRF Analysis, committee member for the Standards Australia Aqua Regia Digest of Soils for Heavy Metals Determination by ICP.

- Further committees that Henry has been involved in are the
- Australian X-Ray Analytical Association,
- Fluoridation of Water Supplies Advisory Committee (Queensland Health)
- and the Fluoridation of Water Supplies Working Group (Queensland Health)

Henry's professional interests include methods of analysis by X-ray Fluorescence Spectrometry, Atomic Absorption Spectroscopy, Inductively Coupled Plasma Atomic Emission Spectroscopy, and Inductively Coupled Plasma Mass Spectrometry with emphasis on matrix correction. His primary focus included speciation studies involving arsenic, mercury, and chromium in complex matrices such as foods including fish, soils, sludges, and sediments. His other interests also include a sampling technique for metals in waters known as diffusive gradient in thin films (DGT). This is a passive sampling technique and offers the major advantage of providing time integrated data.

In the latter part of his career, he was invited to visit China and Vietnam to speak at various training courses in those countries.

During his career he became a NATA assessor, which continued well beyond his retirement. He assessed numerous laboratories throughout Queensland as well as interstate and overseas. These included NATA chemical testing assessments in Papua New Guinea including the Porgera Joint Venture Gold Mining Operations laboratory and the Department of Mines laboratory, Port Moresby.

As Chief Chemist he coordinated and participated in the Inorganic Chemistry laboratory activities which provided high quality analytical work using a wide range of instruments including ICPMS, ICPAES, XRF, GFAAS, FAAS, and wet chemical procedures. Later hyphenated techniques became available such as HPLC-ICPMS. These enabled the development of speciation methods, in particular, arsenic, mercury and chromium species in various matrices.

The work of his area also included analysis of core samples of rocks as part of a complete geological survey of the state of Queensland being carried out by the Department of Mines. The project had been going on since 1936. Unfortunately, only a few years before the completion of the project the executive of the laboratory decided to outsource this work claiming it was not core business of public health. Henry had a sign on the wall in protest, "Cores are our Business."

He participated in multidisciplinary public health emergency response project teams providing advice on sampling procedures, chemical analyses of soils and waters for suspected toxic heavy metal contaminants and the types of testing required for different situations. Some of the projects included toxic waste seepage, arsenic contamination in school grounds, contaminated land sites in public areas, and several mining-related contamination incidents.

He coordinated, supervised, and participated in projects and research and development. Examples include 'Determination of background levels of Bromide in Human Blood' and 'Speciation of Arsenic in Human Urine.'

Before retirement, he was the Chief Chemist of Inorganic Chemistry in FSS. Henry retired from the laboratory in December 2012 after 44 years of service.

MARY HODGE {NEE GRIFFITH} FRACI

Chief Chemist Organic Chemistry 1969-2011

Mary Hodge completed a double major in Biochemistry and Chemistry at Queensland University at the end of 1968. She commenced working at the laboratory in 1969. At the interview she was told she would be working in the forensic area. She was a little concerned about commencing in this area, which involved court work, at an early stage in her career.

She was told by the director that if she was intelligent enough to complete a degree, she was intelligent enough to follow a method. She commenced work in the Forensic Toxicology area and remained there for about 15 months before moving into the Customs area to work with John Foreman. She looks upon this as a very important learning step in her career. The Customs laboratory analysed any product imported into Australia that involved a tariff classification.

So early in her career she gained a broad knowledge of the chemistry of commodities. At the start of her career the analytical work was dominated by wet chemical methods including thin layer chromatography and instrumentation which was devoid of any type of automation. The chemist had to stand by the instrument and inject the samples and use a stopwatch to time and record the peaks on a chart recorder. On occasions the cut-and-weigh method was used to determine the amount of substance. The chart paper was taken from the machine and the peaks manually cut from the chart with a pair of scissors and then weighed. The weight of the paper was directly proportional to the amount of substance present. Later integrators which recorded the area and time of peak were combined with auto injectors to relieve the chemist of the necessity of sitting beside the instrument recording information. Towards the end of her career, instrumentation had reached the stage of processing the data through to a report stage. Through her efforts the instrumentation of the organics section managed to be at the forefront of analytical technology.

This type of automation is a long way from the slide rule which she received when she started at the laboratory. In the early 1970s, pocket calculators started to appear. They were extremely expensive at the time and in one section the calculator was bolted to a large piece of wood to prevent it being misplaced.

In 1975, six chemists were selected to spend a year working in different parts of the laboratory. These chemists were seen as future managers. Mary was selected as one of these chemists. During this time, they gained an overview of the work of other departments. In 1983 she took a short break from work to commence her family. On her return to the lab in 1984, she moved into the pesticide residue laboratories. She remained in this area for the rest of her career. The section's name was changed to Food Quality in 1995 and later to the Organics section.

During this time residue chemistry changed to become an important political issue. Mary worked on the Condamine/Ballone project which looked at contamination in that river system and the Emerald Leukemia cluster project which looked at pesticide spray drift from cotton farms. She was later the laboratory representative on the committee looking at the quality of recycled water including water extracted from sewage and desalinated water.

Part of this project involved analysis of a very wide range of pharmaceuticals, cleaning products, personal care products, endocrine disruptors, pesticides, and herbicides. The laboratory under her direction developed methods to efficiently analyse samples for all these different kinds of products. This involved the purchase of new equipment and the training of staff.

During her period in charge of the organics section the government introduced a more competitive basis for analytical work. This meant that the laboratory had to compete for tenders. She spent many weekends writing tender proposals to ensure the laboratory had sufficient work. The laboratory was the highest money earning laboratory in the organisation. Over countless years the funds earned by Mary's section saved the laboratory's budget.

As Chief Chemist of the Organics section, Mary was recognised as an exemplar and mentor for other women chemists. She was asked on one occasion to address a women's chemists' breakfast. She made the point that hard work and working with and supporting a good team was a requirement for anyone to rise to higher levels in the organisation.

Mary helped organise several training workshops for Chinese chemists at the laboratory. She organised two national conferences for the pesticide residue chemists. In 2011, Mary was Chairperson of the 23rd Conference of Residue Chemists held in Brisbane. This provided an opportunity for the residue chemists in this area of Australia to build networks. In recognition of her work and status, she was elected a

Fellow of the Royal Australian Chemical Institute. She retired in 2011 after 42 years of service to the laboratory.

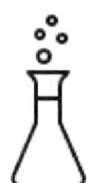

FIRE DUE TO MAGNESIUM PHOSPHIDE FUMIGATION

The local council reported several fires in waste disposal facilities. The source of the fires was packages of magnesium phosphide used to fumigate shipping containers. The magnesium phosphide reacts with moisture and produces spontaneously flammable toxic phosphene gas. A decision was made to collect the packages and immerse them in a 200L drum filled with water. This worked on several occasions. On one occasion the drum was only half filled. An explosive atmosphere developed in the space and a loud explosion occurred. Fortunately, the emergency responders were wearing full protective equipment.

SMILE OR A FROWN

Up until the 1970s the women's sexual health clinic was in the laboratory grounds. The lab store manager said that during the second world war, the American troops used to wait outside the back gate to see if the Ladies of the night came out of the clinic with a smile or a frown.

POISONED PHARMACEUTICALS

Graeme White assisted police in several high-profile cases including one, involving the 'random' adulteration of paracetamol with strychnine, connected to an alleged attempted homicide. He worked on site with police in the scanning of a suspect's house where traces of strychnine were found in several locations, using an ion mobility spectrometer borrowed from the Borallon Correctional Facility. On analysis at the laboratory, he detected 10-50 nanogram levels in a few samples.

GRAEME WHITE
Principal Adviser, Forensic Sciences 1979-2011

Upon completion of studies at Sandgate High School, Graeme commenced work at Usher's Paints at Geebung as an Industrial Chemist. He attended QIT part-time for six years to complete his Degree in Applied Science (Industrial Chemistry) and followed that with a Graduate Diploma in Business Administration. After three years he left Usher's and commenced work at Nightingale Chemicals at Eagle Farm. Here he gained a wide range of experience in the industrial synthesis of raw chemicals as well as many commercial and industrial products including cleaning chemicals, sanitizers, sealants, adhesives, polishes, and aerosol deodorisers. He was around when the only adhesive the aviation authorities would approve for the construction of wooden plane components was casein, and telephone books were bound using animal glue made at Eagle Farm. He was eventually Chief Chemist at the Brisbane factory of Nightingale Chemicals.

He was successful in applying for a chemist position at the GCL in 1979. He started in the foods laboratory where he was appointed a State Analyst under the *Health Act* and then moved to Forensic Chemistry in 1986 just as the new *Drugs Misuse Act* was gazetted, under which he was appointed an Analyst. At that time, the *Drugs Misuse Act* prescribed a mandated life sentence for drug possession involving, for example >2 grams of heroin or >0.004 grams of LSD. This certainly focused the mind when reporting results of 2.05 grams of heroin. At the time Forensic Chemistry and Forensic Toxicology were a combined unit with one supervising chemist, Geoff Rynja.

After a while, the next section's Supervising Scientist, Wayne Pease, asked Graeme if he would like to work in physical evidence saying at the time: 'I have to warn you it is the pits'. Wayne later recounted that the then Director, Don Lecky asked him how Graeme was handling physical evidence, to which Wayne replied, 'he is taking to it like a duck to water'. Wayne had a way with words, in lobbying Des Connell for additional forensic staff, only available from the environmental laboratory, he coined the phrase 'environmental BS samples' as a means of comparing his view of the relative importance of environmental samples compared with forensic samples. In physical evidence Graeme did a lot of work on oil spill identification to determine the source. His thesis on oil spill analysis gained him a Senior Chemist position.

Following Wayne's departure, Graeme took over as Supervising Scientist and undertook significant initiatives improving productivity, to cope with ever increasing demand for forensic services. Quality systems (ISO 9001) implementation started in 1994 and in 1998 his laboratory gained NATA forensic laboratory accreditation. The laboratory's instruments were interfaced to the Internal Forensic Chemistry computer network built with major input from Chemist Methsiri Edirisinghe. Data flowed from balances to instruments and results were automatically processed and delivered directly to chemist's workstations. Initially some junior chemists were tasked with processing the samples through the automated instruments.

The task of receiving cases, which had always fallen to chemists on a roster basis, was now performed by a newly created exhibit officer position in a specially built Forensic Chemistry Property Point. Elaine Slapp was the first Property Officer on site and gave functional, intelligent advice in the

development of the property point. At the time Graeme's section had to forgo one position within the laboratory to get a Property Officer.

By 1994 clandestine laboratory seizures had started to increase and rapidly grew. A separate subsection within Forensic Chemistry was created to develop the necessary skills and knowledge to conduct analysis and to assist police at the site of these clandestine laboratories. This section also managed the substantial accumulation of highly toxic evidence up to and after the closure of a case. Part of that solution involved the QPS building their property point on the site abutting Forensic Chemistry.

To manage the growing problem of police demand for services and the resources required to provide those services the then manager of forensic sciences, Ron Biltoft, initiated the development of the Inter Departmental Standing Committee for Forensic Sciences. The committee consisted of high-level decision makers from Queensland Health, the Department of Justice, QPS and Legal Aid. Some members would participate for specific issue meetings, e.g., the State Coroner participated in issues relating to the timely provision of reports from

Forensic Pathology and Toxicology, which were pivotal to the coronial process of finding cause of death and the release of deceased to loved ones. The Chief Magistrate would attend when matters arose in connection with the structure and content of analysts' reports to the courts. Graeme provided much input to this committee and was also the secretary to the committee for many years.

When Ron Biltoft, the Manager of Forensic Sciences, retired in 1999, Graeme was appointed acting manager until a new manager was appointed in 2000. With the appointment of a new manager and resultant restructuring in 2000, he was appointed into the new position of Scientific Manager to enable retention of NATA forensic laboratory accreditation which required the manager to have five years' experience working in a forensic laboratory. This was not considered as part of the selection process. The appointed manager did not have any forensic experience. Following a tumultuous period, in 2002 the positions of Manager Forensic Sciences and Scientific Manager Forensic Sciences were both made surplus to needs and both incumbents were made redundant. Graeme applied for the position of Scientific Advisor and was appointed in 2003. In that position he focused on the development of forensic research, staff development and education.

Since 1993 staff had been assisting Associate Professor Dennis Burns in the development of Griffith University's forensic sciences courses. Graeme provided lectures at Bond University, Griffith

University and QUT. He was appointed to the honorary positions of Adjunct Associate Professor at Bond and Griffith universities.

In 2006, following consultation with Maree Storer of the Department of Employment, Economic Development, and Innovation (DEEDI), he approached Professor Burns with a proposition to seek grant funds from DEEDI for the development of forensic research in Queensland to assist with staff development and to make advances in capability and productivity. With the approval and assistance of the then Director, Greg Shaw, Dennis and Graeme successfully received a Smart State grant for $100,000, which was leveraged to $378,000 through the participation of recruited partner organisations. Using this 2008 funding he and Dennis won a further $2m grant which was leveraged with participating partners to $6.8m. This funding developed the AFFIN which ran seven significant demonstrator research projects across Australia and was an alliance between 19 national and international organisations. The most productive project is still running and is run by Professor Jochen Mueller of UQ. The project initially looked at the detection of illicit drugs in sewage flow streams coming from specific target locations.

Graeme retired in 2011 after 32 years' service to the laboratory.

DIOXIN

Dioxin is a highly carcinogenic chemical. In the 1960s, the laboratory purchased a standard of the most dangerous isomer of dioxin. The director of the time decided that it was too dangerous to keep as the laboratory did not have the facilities to adequately protect the staff. The sample was sealed inside a metal cylinder and then cast into a block of concrete. A metal sign was placed on the outside of the block. For the next 40 years, this block of concrete continued to be a source of concern. When the William Street labs were being demolished it was found in the vault and returned to the lab. On one occasion it was suggested that it be buried under the foundations of the new building. However, taking a long-term view this did not occur and eventually in about 2010 the sample was disposed of by a commercial company.

STEWART CARSWELL FRACI
Chief Chemist 1992-present

Stewart Carswell completed a Master of Science in analytical chemistry from QUT in 1991 under Serge Kokot, before starting at the GCL in June 1992. He started in the Manufactured Products section (later renamed Investigative Chemistry) where there was a chemist vacancy due to recent loss of a staff member. It was a great grounding for a fresh graduate as the section dealt with a wide range of unusual samples and analyses. Wet chemistry was still a major part of the analysis, and the staff were very experienced. One of the first jobs was to deal with the backlog of dog baits. Generally, the samples were unpleasant either rotten meat, vomit or worse. Testing involved FTIR, GCMS, wet chemistry and spot tests. An unusual sample was a molasses and grain mix laced with white phosphorus. The evaluation of the state-wide government laundry contract also tested his analytical techniques, wet chemistry knowledge and method development skills.

While in Manufactured Products he joined the Response Advice for Chemical Emergencies (RACE) team (later Queensland Fire and Rescue - Scientific Branch) as an on-call officer to assist the fire brigade deal with chemical emergencies. He remained as a volunteer for over 20 years. The skills learned, the training and the exposure to situations outside of the normal laboratory environment expanded his advisory and scientific skills. The scale of the work changed. Instead of dealing with a beaker of acid, you're suddenly dealing with a 20,000 L spill of concentrated acid running into the local creek. This role was highly valued by the client and this appreciation added to his enjoyment of working at the GCL/FSS. He received an Australia Day award for this work.

He was identified by Gary Golding and Mary Hodge as a potential future leader, so needed to broaden his knowledge of the general work of the laboratory. After 6 years in Manufactured Products, he was transferred to the Mining (later Inorganics lab) for a short period, where he did iron oxide titrations, HF digests, metals by AAS and total organic carbon. It was more routine than Manufactured Products and when asked to move to the Environmental Waters section he decided to take the opportunity. Liquid-liquid extractions, GC-FID, GC-ECD and GCMS analysis became a normal day to day activity.

Environmental Waters eventually became part of the Organic Chemistry laboratory in 2002, and he moved upstairs to level 2. While the analysis of PAH, Phenols,

TPH and THM/HAA were still the focus of his work he was exposed to pesticide residues analysis. The work changed with the development of other extraction techniques including Solid Phase Extraction. New residues were analysed, more sensitive, faster instruments were purchased.

During this time Mary Hodge was his supervisor. In 2012, he became Chief Chemist of the Organic Chemistry lab. This was a difficult time, as described elsewhere, as staff cuts were harsh and not well directed with the loss of many experienced staff. It was a baptism of fire and a very distressing time for him and his staff. They got through it and have continued to develop new techniques and utilise new instrumentation including high resolution mass spectrometry which moved out of the research lab and has become more routine. The chief chemist role is now in charge of a wider range of labs than previously and has moved to a more bureaucratic management role. He says, "I no longer own a lab coat." He added that "The changes over the 30 years at the lab have been dramatic but the thing that makes it work are the people. Being able to achieve what we do, to be considered a "go to" lab for our clients and considered a great place to work (by most of the staff) and to help staff build competence is what keeps me going".

Like many leaders at the laboratory throughout the past 100 years, he has been involved in the Royal Australian Chemical Institute (RACI) especially the Analytical and Environmental group, which he joined as a committee member (secretary) in 1992. He is still a member. He has continued in various roles, including group Chair and Queensland Branch President. He was also the joint Chair of the 2004 Analytical, Environment and Electrochemistry national conference and was elected a Fellow of the RACI in 2017. Recently, he has been mentoring post-graduate students in their final year of study, not in their field of research, but to aid in the transition to the "real" world with job application and interview skills. The RACI has opened many doors and opportunities, and he'd encourage others to consider joining and be involved.

Gary Golding was his first supervisor and one piece of advice that he still remembers is "you can have a result, quick, cheap or right – pick one." This relates to the project management triangle of quality, time, and price. All work falls within that triangle and is a compromise of these three parameters.

One final comment Stewart made, "When I first started, I couldn't understand how all these old people had spent their entire careers at the lab and I thought, no way! I was planning to spend a few years, get some experience and move onto a better job. As I've stayed, I've come to realise that firstly, they weren't that old and secondly, this was a great place to work with different labs to work in and different problems to solve that couldn't be compared with other workplaces. The variety of work has kept me here for close to 30 years. "

JOHN BROWNLIE HENDERSON FRACI

Government Analyst/Director 1893-1936

Longest serving Director

This article was published in Australian Dictionary of Biography, Volume 6, (MUP), 1983 by H. J. Gibbney.

John Brownlie Henderson (1869-1950), Analytical Chemist, was born on 29 April 1869 at Barrhead, near Glasgow, Scotland, son of Robert Henderson, Cashier, and his wife Jane, nee Kinloch. After primary education at boarding schools, he attended Allan Glen's Technical School and in 1887-88 Anderson's University, Glasgow, now the University of Strathclyde. He became Research Assistant to Professor William Dittmar, particularly in a complex investigation of the gravimetric composition of water which won the Graham gold medal for research.

Following his parents to Queensland in 1891, John Brownlie Henderson became Science Master at Brisbane Grammar School and in 1893 Government Analyst. He returned to Britain in 1899 and married Jean Susan McKeown at Leenane, Connemara, Ireland, they had four sons. As an Analyst Henderson undertook chemical work for all state departments except the Department of Agriculture, and eventually for federal departments in Brisbane. With J. C. Brünnich, his agricultural counterpart, he had a standing demarcation dispute. The laboratory contained much equipment designed by him.

Ivo Henderson at his retirement commented that the previous John Brownlie Henderson, was a bit odd, and liked to paint things in the lab black. Henderson and his staff undertook a constant flow of routine tasks, punctuated by such high points as the testing of the first significant Australian oil discovered at Roma.

Something of an egotist, he spoke always of 'my laboratory.' Asked to draft the Explosives Act of 1906, he became Chief Inspector under the Act in 1907 and supervised the creation of magazines all over Queensland. An explosives area at Narangba was named after him in 1951. During the First World War, he was a member of the State Munitions Committee which arranged the dispatch of 600 workers to Britain and chaired the committee which controlled all explosives in the state.

An enthusiast for education, John Brownlie Henderson was active both in the Sydney University Extension Committee and the movement which finally secured a university for Brisbane. In the first of three terms as president of the Royal Society of Queensland, he devoted his presidential address to a review of Queensland education and a plea for a university. Nominated to the first university senate, he resigned only on retirement in 1936.

HENDERSON'S WORK

The annual reports from 1927 to 1938 give some information on the topics of the day. Lead in soda water was a recuring theme. Lead levels of up to 1700ppm were detected. This was leached out of the solder used in the construction of the carbonator. A similar investigation was carried out in the early 2000s at the laboratory on the heavy metal content of water from instant boilers used in canteens.

Other reported cases included.

- Formaldehyde in canned and smoked fish
- Composition of Queensland bread
- Dirt in Milk
- Raspberry cordial containing no fruit juice
- Adulterated vinegar
- Beer with excess sulphur dioxide
- Glass fragments in cordials and bottled fruit
- Sponge cakes containing boric acid
- Lead Arsenate in vegetables and fruit
- Lead in paint, crayons, and toys
- Paraffin oils on currants and raisins
- Vitamin tablets and cod liver oil deficient in vitamin A and D
- Air spaces in bread

- Copper in tinned peas

John Brownlie Henderson chaired the Chemistry section of the 1904 meeting at Dunedin of the Australasian Association for the Advancement of Science. In 1917 he was a founder (fellow 1918) of the Royal Australian Chemical Institute and was its first Queensland president.

Appointed to the State committee of the Advisory Council for Science and Industry in 1916, he acted in 1917 for a chairman on war service and joined ex officio the national executive committee. He chaired the provisional State Advisory Board of the Institute of Science and Industry from March 1921 and became a member of the state committee of the Council for Scientific and Industrial Research in November 1928, resigning in 1949 when it was reconstituted. John Brownlie Henderson was an executive member of the Queensland branch of the Australian Red Cross Society from 1914-18. Later, as an Officer of the Boy Scouts' Association, he acted as Providore for a world jamboree in Brisbane. Tall and thin with strong cheekbones and a 'handlebar' moustache, John Brownlie Henderson was an excellent lecturer despite a tendency to verbosity and over-emphasis of sibilants. He was scientifically cautious -when asked for forensic evidence: police officers called him 'the silent witness' because he refused to venture opinions and stuck to facts. He died at Annerley on 19 October 1950.

THE TOP AND BOTTOM OF ORGANISATIONS THINK DIFFERENTLY

Quantifying small amounts of drugs was taking up a lot of time and resources. The total amount of powder is often less than the lowest statutory amount. There appeared to be little point quantifying the amount of drug in small amounts of the powder. The chief magistrate agreed that very small quantities of powder need not be quantified. Unfortunately, the sitting magistrates wanted to know the quantity of drug found on the person to determine the penalty. To overcome this problem, it was decided to send out reports stating the identity of the drug and that the sample could be quantified if requested. As mentioned previously, attempts to stop analysing dog baits was unsuccessful because upper levels focused on the likelihood of a successful prosecution and other levels focused on safety.

EXHIBITS TAMPERED WITH.

The laboratory routinely destroyed drug exhibits for the police. The analyst who was on duty recognised an exhibit being returned for destruction. The original sample he had analysed was a pink rock heroin. The sample was now a white powder. This shows the advantages of staff who are observant and carry out all stages of the process.

OTHER MANAGERS

The following extracts of staff member's personal files show the development of the careers of various managers at the laboratory.

JOHN {JACK} ADAMSON

Senior Analyst 1913 - 1963

Extract from personnel records of John Adamson:
1913 Shown in 1913 staff photo

1915 - 1919 - Active service in Australian Imperial Forces (AIF)

1925 Appointed Analyst, Government Chemical Laboratory

1947 Appointed Senior Analyst, Division I, and Inspector of Explosives 1951 Queensland Government departmental restructure - Lab work for Dept of Commerce and Agriculture now carried out at Customs Section of Laboratory under direction of Mr Adamson

Photographs from 1913 show Adamson worked at the laboratory prior to enlisting for military service. The war disrupted his studies for a degree in science, however he was still promoted to analyst presumably based upon experience. The RACI recognised the role of experience in the training of chemists. The title Chartered Chemist (CCHEM) was often awarded to people without a full degree in chemistry but with extensive experience in a particular field of chemistry.

WILLIAM {BILL} NEVILLE CARVOSSO
Chief Chemist Waters 1946-1976

Extract from Bill Carvosso's personnel records

1929-1933 employed by Public Analyst and Assayer (Mr H C Oakes)

1934-1947 employed at The Brisbane Gas Company (Gasworks, Newstead) as Assistant Chemist

1947 Bill Carvosso acceptance of appointment to Analyst, Government Chemical Laboratory, to commence duty 27 Oct 1947

1954 Appointment as State Analyst under the Health Act

1954 From SB Watkins to DG Health and Medical Services re: approval for WN Carvosso (Senior Water Analyst) to join a research party investigating coral death at the Great Barrier Reef

1955 Promotion to Chemist, Division I (Department of Health and Home Affairs).

1959 From SB Watkins to DG Health and Medical Services — support for request by WN Carvosso [unknown request type] Operating out of Colchester Street and head of the waters section

1960 Recommendation letter that WN Carvosso be appointed Research Officer

1962 Appointed to Senior Chemist, Government Chemical Laboratory, Dept of Health and Home Affairs.

1965 Notification of appointment to Chief Chemist, GCL, Department of Health1976 Direction to redesignate positions within the GCL

1976 Appointment as Supervising Chemist, GCL, Department of Health

It is interesting to note that the laboratory was involved in determining the cause of coral death on the Great Barrier Reef back in 1954.

HOWARD COUPER
Deputy Director 1946-1979

Howard Couper oversaw the mining and industrial hygiene laboratory also known as the Ores Lab. He delighted in calling new chemists "bucket chemists" if he saw them manipulating large quantities of reagents. He was the last of the chemists to carry out their work using wet chemistry (titrations, gravimetric analysis, electro chemistry), instead of instruments. The next generation of chemists were more analytical physicists than analytical chemists, applying spectroscopy in all its forms to carry out the analysis. He was deeply involved in the investigation of the explosion at the Box Flat coal mine near Ipswich.

Chemists spent a lot of time travelling the State and in factories testing air. Work at that time also included,

- Analysis of ore samples from Cloncurry assay office for gold, silver and copper, antimony, arsenic
- Testing coal samples for the Coal Board
- Mine dust sampling
- Testing diesel exhaust from vehicles used in underground coal mines
- Lead in air
- PVC monomer associated with PVC pipe manufacture Respirable Silica in air
- Asbestos fibres in factories producing fibro (asbestos cement sheeting for housing construction)
- Trouble shooting problems e.g., cause of corrosion in air conditioning units

- The lab periodically received gold fillings from the teeth of deceased persons. This was purified and the gold donated to a charity.
- Firing of clay samples to determine their use for brick making, expanded clay products, pottery

Extract from Howard Couper's personnel records:

1946 Appointed Temporary Attendant, Government Chemical Laboratory

1946 Appointed Attendant on probation

1952 Application for the position of Government analyst has fulfilled the requirements for admission as an associate in the Royal Australian Chemical Institute

1952 Appointed Analyst

1958 Appointed Chemist, Division 2

1961 From Under to UQ Registrar — permission for Couper to give 28 three-hour lectures on assaying during the academic year.

1965 List of teeth forwarded to Govt analyst

1959 Appointed Chemist, Division I

1969 Appointed Senior Chemist

1966 Appointed part-time Lecturer in Production Metallurgy I and Metallurgy II at UQ

1968 Appointed part-time Lecturer in Chemistry at QIT

1976 Appointed Supervising Chemist

1979 Appointed Assistant Director, Government Chemical Laboratory

KEITH DEASY
Deputy director 1945-1984

Extract from personnel records:

1940 Cadet on probation, Survey Office

1940 Cadet, Survey Office

1945 Assistant Draughtsman, Survey Office

1945 Assistant to Analysts, Government Chemical Laboratory

Resigned as Assistant to Analysts, took temporary position as Demonstrator, Chemistry Department, The University of Queensland

1949 Appointed Analyst, Government Chemical Laboratory

1957 Appointed Chemist, Division I

1957 SB Watkins strongly recommends that Mr Keith Deasy be promoted to Chemist

Division I. "Mr Deasy is expert in testing chemical balances and other delicate laboratory equipment and in all his work he brings to bear a high degree of intelligence backed by sound chemical knowledge."

1962 Appointed Senior Chemist

1965 Appointed Inspector of Explosives 1965 Appointed Chief Chemist

1968 Nominated to attend a computer training course

1969 Ivo Henderson seeking approval for Mr Deasy to attend Conference on X-ray Spectrometry at Australian National University between 9 — 13 February 1970. Registration fee $20.

1970 Appointed State Analyst

1970 Report on conference proceedings. The conference was successful and resulted in Mr Deasy purchasing a number of textbooks for the library as well as a GCL computer.

1973 Letter from State Fire Services Council Queensland requesting Mr Deasy to give a lecture on "Dust Explosions" on Wednesday 1st August 1973 from 9am — 11.15am.

1975 Made assessor for National Association of Testing Authorities

1975 Appointed Deputy Director, Government Chemical Laboratory

1977 Relieved of appointment as Deputy Director. Appointed Assistant Director 1979 Appointed Deputy Director

1984 Retired

Keith Deasy was responsible for the modernising of the laboratory's instruments. He set up the spectroscopy laboratory as a centre for instrumental analysis. He developed Xray Fluorescence (XRF) capability, expanded Atomic Absorption Spectroscopy (AAS) techniques, Gas chromatography (GC), emission spectroscopy. These instruments represented a major investment by the state, and they were all used for many years. The emission spectrograph, however, was soon superseded by newer technologies and was redesigned into an inductively coupled plasma spectrometer which carried out the same elemental analyses, but on liquid samples. This specialised instrument section provided services to the other sections of the laboratory. He had the good fortune of being the longest living type 1 diabetic in Australia having the good fortune of being diagnosed soon after insulin became available.

CABINET SUBMISSIONS

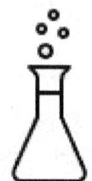

Prior to leaving, Ron Biltoft gave Gary Golding some worthwhile advice. If you want more resources, put together a cabinet submission. Don't just ask the department. Gary put this into action. However, to reduce spending cabinet submissions were suspended. Once the embargo was over Greg Shaw put together a submission that resulted in some good outcomes for the laboratory generally and chemistry in particular.

INSTRUMENT CONVERSION

In the early 1970's the lab purchased an emission spectrograph. The emission spectrograph utilised an electrical spark to excite the elements and caused the emission of light. This light was separated into its wavelengths and the spectra recorded on a 35 mm film. The film was developed, and the density of the spectral lines measured on a densitometer. A rather cumbersome process. Later the instrument was converted to an inductively coupled plasma (ICP)

spectrometer. It used a photomultiplier which obviated the need for photographic film. One of the chemists, David Grantham, developed a process to measure the spectral line intensity and correct for the background emission intensity at a nearby wavelength using a silica window with a second slit at a slightly offset wavelength. Two large black lines were attached to a refrigerator door and photographed. This photograph of the two lines was then reduced to the right size (width about 5 microns) on the film. The image was transferred onto the silica window coated with a light sensitive resin. The unexposed resin was washed off and the window aluminised to produce a mirror with the two resin lines. The resin was then dissolved off to leave two lines of the right dimensions for the measurement.

METHSIRI EDIRISINGHE

Chief Chemist, Forensic Chemistry 1992-2017

Methsiri Edirisinghe obtained his BSc (Hons) degree from the then University of Ceylon and MSc degree from the UNSW. He served as a State Analyst from 1974 to the end of 1990 in the Government Analyst's Department of Sri Lanka. He worked in the field of Forensic Science covering many disciplines such as Serology, Toxicology, Drugs of Abuse, Firearms and Trace Evidence. At the time of leaving the Department he was the senior analyst in the Drugs of Abuse area. During his tenure as a State Analyst, he spent 10 months (1978) in the Swedish National Forensic Science Laboratory (SKL), Linkoping, Sweden, and about two and half years (1983-1986) in UNSW, Sydney reading for MSc degree by research. Although he was a trainee scientist at the Swedish laboratory, he participated in a research project related to clandestine manufacture of methamphetamine. He was a co-author of a paper published from the findings of this research in a peer reviewed journal. The Swedish laboratory arranged for him to visit the Metropolitan Police Laboratory in London and the University of Strathclyde during his stay. He was also involved in providing lectures at the Sri Lanka Police Training Academy. Prior to the resignation from his position in Sri Lanka he took the leading role in designing and equipping an analytical laboratory at the invitation of the National Dangerous Drug Control Board of Sri Lanka.

He migrated to Australia in January 1991. The advice he received from the many people he knew in Sydney was, to expect a long unemployment as Australia was in recession. Against all odds, he secured a QA scientist position at Peptide Technology Ltd in March 1991. His love for forensic science never faded. When an opportunity opened (1992) at the GCL, for a position of an analyst in the Forensic Chemistry team, he applied. As a result, he joined the Forensic Chemistry team of the GCL as an analyst in June 1992. He worked in all three groups of the team but spent the bulk of the time in the illicit drug group. The team leader at the time was Graeme White and he was very unhappy with the ageing computers in the instrument room as the technical issues were hampering the efficient data acquisition and analysis. His attempts to replace them through the laboratory's IT unit were fruitless as the instrument computers did not fall within the domain of IT.

Methsiri, by that time, had developed good skills in building and troubleshooting PCs and suggested to Graeme White that he could build new PCs if parts could be procured. The parts were procured utilizing the team's budget. The instrument PCs were successfully renewed, and Graeme soon conceptualised a mini network of computers so analysts could access the data acquired by the instruments from the laboratory desk. A program was also developed to automate the data analysis, thereby minimising transcription errors. Graeme shared the network idea with Methsiri. Methsiri apologetically declined to take up the task citing that he had no knowledge or the experience as networking is a totally different ball game to assembling or troubleshooting PCs.

A week or so later Graeme tossed a textbook on to Methsiri's desk and walked away saying "I am sure you can do it." The book was titled "Networking for Dummies." A few weeks later Methsiri was able to obtain necessary cables, plugs and other tools from the IT unit free of charge. Graeme and Methsiri spent several weekends in the lab, at times in the ceiling cavity, and in the end the "FChem" network was born. This network survived even after the introduction of two new laboratory data management systems.

Methsiri volunteered to support Graeme in getting the laboratory prepared for its inaugural NATA accreditation. Both spent many weekends and nights getting the documentation ready. The efforts were very rewarding.

In 1995 GCL ceased to exist and in came QHSS. Nothing changed overnight except the name. Forensic Chemistry was struggling under the ever-increasing inflow of work, and it continued to do so. Achieving NATA accreditation was an added impact on workload despite it being a necessity. Upper management kept telling the middle managers that the process had to be cost neutral. While this seemed to be an idiotic statement the message was to prepare yourselves for accreditation, but don't ask for resources.

In 1999 Graeme White moved on to become the Scientific Advisor to the new Manager, Forensic Sciences and the position of team leader, Forensic Chemistry fell vacant. The new Manager, to save her budget, was not very keen on filling that vacancy immediately and instead opted to appoint interested parties in an acting capacity. Methsiri could recall various instances where the management implied that the staff contributed heavily to the problematic issues prevailing at the time. The reality was that the upper management did not either understand the message that the team leaders were trying to deliver or chose not to understand. However, the Forensic Chemistry staff was vocal in showing their disagreement to prolonging the acting role of the team leader. The vacancy was advertised and was filled in 2000. Methsiri was the successful candidate.

Despite being under-resourced (staff wise) Methsiri was keen to find ways to improve analytical processes. Methsiri was a member of the Specialist Advisory Group (SAG) for forensic chemistry. The SAG was a group comprising of the Forensic Labs in Australia and New Zealand who met regularly to improve systems. Through the SAG he became aware of the work being carried out by the Senior Working Group for the Analysis of Seized Drugs (SWGDRUG – based in USA). It was developing analytical protocols for Forensic Labs. Forensic Chemistry was the first laboratory in Australia to adopt the new acceptance criteria in drug analysis approved by the SWGDRUG as the new standard. It removed over-testing thereby leading to efficiencies. As the results are subject to scrutiny in a Court of Law, these standards provided strong support for the analysts to defend the amount of testing carried out on exhibits.

Methsiri was also skilled in acquiring new technology. LCMSMS and FTIR were among them. He also brought in liquid handling technology automating sample dilution further enhancing accuracy and efficiency.

Around 2003 a decision was taken to add more staff resources to Forensics. Each team leader was to propose their requirements and justify their claims at a meeting with the CEO. Methsiri asked for four senior chemist positions, (none existed at the time) two chemist and two technician positions. During a tea break of the meeting Methsiri was told by many attendees that the request made was way too optimistic and there was no way he could justify four senior chemist positions. Methsiri provided his arguments for it and the meeting concluded with decisions to be made in due course. Methsiri shared this information with the staff and the staff were very pleased with what he had asked for. However, the consensus was that it's unlikely that we would get more than two senior positions, which seemed to be the norm. One made the comment that if we get all four senior positions a statue of Methsiri should be erected in the lab. A few weeks later the approval came from the head office for all Methsiri has asked for including the four senior chemist positions. However, the erection of the statue of Methsiri did not proceed. (Editorial comment: Unfortunately, Methsiri did not formally submit a request for a statue in the submission.)

Methsiri worked closely with Graeme and Griffith University in developing course content for a MSc course in Forensic Science particularly in the disciplines covered by the Forensic Chemistry team. He was also a lecturer for the course as well.

He held the chair of the SAG for two years and continued to work in close collaboration with other Forensic laboratories in Australia and New Zealand. He joined NATA on invitation as a technical assessor (voluntary position). Prof Olaf Drummer invited him to be a reviewer of research articles on illicit drugs submitted for publication in Forensic Science International. He functioned as a reviewer from 2010 until 2017. He encouraged staff to be involved in research and engage with external agencies for mutual benefits. He has been a co-author of a few papers published in peer reviewed journals. A couple of papers were in collaboration with ENTOX on illicit drugs in wastewater.

Methsiri held the view that a cohesive team is a happy team. For a team to become cohesive, he believed that open communication and the freedom to express a point of view are of paramount importance. This was not a common concept in the larger multidisciplinary organisation. He believed that he achieved that to a very satisfactory level and received great support from the staff as well in this endeavor. This was seen by some as "Methsiri had no control of his staff at meetings, and they would keep talking about whatever came to mind." The impartial culture ingrained in the chemistry group meant that they were free with their opinions and felt it was for the good of the organisation that they express them rather than let senior management decide without examining all options.

This free expression of opinions in a political environment led to the upper management thinking all issues or problems were created or caused by the chemists. One of many occasions which demonstrated an intolerance of criticism was seen at a large gathering of chemists. One of the chemists complained about a lack of forensic resources. A high-level manager took over the rostrum and proceeded to berate this person saying, a lot of resources had already been supplied. One of the chief chemists went to the rostrum and whispered that the person raising the resources issue was from the South Australia forensic laboratory, which is in a different jurisdiction and did not fall under this manager's control. How senior managers developed this immediate defensive attitude to chemist's thoughts is unclear but the term "white anting of chemists generally by other political operators within the laboratory" comes to mind. However, in truth, given the right tools and management environment the chemists perform miracles by solving problems. For example, building their own computer network that actually worked.

2012 saw a big cull in the public sector. Methsiri had a series of meetings with the Senior Director at the time, Greg Shaw, to discuss the staff cuts. Greg Shaw agreed that Forensic Chemistry staff numbers should not be reduced. Methsiri was very happy and informed the staff of the Senior Director's decision. The consultants that made up the outsourced committee enforcing staff reduction were already in FSS. A few days later Greg Shaw went on annual leave and Methsiri was on recreation leave and the committee looked at Forensic Chemistry and demanded 6 positions to be axed. Methsiri came back to the laboratory and attempted to reverse the decision citing the meeting with the Senior Director. Everything fell on deaf ears and the acting Senior Director did not want to get involved. The CEO's phone went to message bank repeatedly. Forensic Chemistry had to lose six staff. Methsiri said, "It was like someone handing me a loaded gun and ordering me to shoot at my own family." Again, the team's cohesiveness came into play. Methsiri managed to lose six staff members with minimal pain. Only one member had to go without consent and the rest took up "voluntary" retirement. What happened two years later exposed the stupidity of the decision to axe 6 positions. Forensic Chemistry had to employ 8 staff on a temporary basis to cope with workloads. These positions were later made permanent. Methsiri considers himself to be very fortunate to have a team of capable, intelligent, dedicated and above all, friendly professionals to work with while at the QHFSS. Methsiri retired in 2017 after 25 years' service to the laboratory. (Perhaps sufficient time to warrant a statue).

SOME THINGS DON'T CHANGE.

The 1947-48 annual report mentioned the lead in paint problem on old buildings, struvite being mistaken for glass in canned fish, excess coal tar dyes in lollies and even the neglected state of the laboratory library. Other things are rarely seen, e.g. phosphorescent lobster due to bacterial contamination.

DR PETER CULSHAW FCS

Chief Chemist of Forensic Chemistry 2016 -

Peter obtained his BSc (Hons) from what is now Sheffield Hallam University in the UK. This was followed by a period within the UK Scientific Civil Service as a Scientific Officer working on the forensic analysis of explosives. Peter then returned to university to do further chemistry research for his PhD at the University of St Andrews, Scotland,

Peter relocated to Australia in early 1991 when he started Post-Doctoral Research at the Research School of Chemistry, Australian National University, Canberra. Peter then spent the new few years working within CSIRO on novel biosensor designs before moving to Queensland Health Forensic & Scientific Services and Forensic Chemistry in 1998.

For a large part of Peter's time in Forensic Chemistry, his area of expertise was the analysis of clandestine drug laboratory cases. This was a very interesting area as it pulled together both analytical and organic chemistry and, apart from the analytical results, required a lot of opinion- based evidence which often in the earlier days required giving evidence frequently in court.

Over these years Peter has seen the change in laboratories and instruments transition from large electronic devices with multiple gauges, dials, and chart recorders, to more compact bench-top featureless systems controlled via computer, and the move away from the lab bench to the office desk for the processing and interpretation of data rather than spending a lot of time in a white coat with test tubes.

Critical to the success of forensic science is quality assurance. Peter has been a NATA technical assessor for over 20 years. This has enabled him to visit most of the forensic labs across Australia. The real benefit of this is the ability to share best practice and new ideas within the forensic community so we are always striving to be the best we can.

Peter has also maintained a close working relationship with academia, especially Griffith University where he has been a sessional lecturer in forensic science-based subjects for over 20 years and was recently appointed Adjunct Associate Professor in the School of Environment and Science.

Scientifically the last 6-7 years have been a very interesting and challenging time for the whole team.

2020 ONWARDS - COVID

COVID changed everything in early 2020. As a lab, we introduced a number of measures to reduce transmission risks including increased separation and isolating, mask wearing and a split shift so that we had morning, and afternoon shifts that remained separate so that we would always have some staff available in the event of an outbreak. We did very well, and no outbreaks occurred, staff were vaccinated and only after borders and other public measurers were relaxed did the first cases come through the lab, but by then we were well prepared and have not had any adverse issues to date.

Teams Meetings online have become the new norm, which has pros and cons. The annual Drug Specialist Advisory Group (SAG) meetings have not occurred in person for the past few years. It is unfortunate not being able to meet up with our interstate colleague's face to face as there is a definite loss in discussions and sharing of information that has been of real benefit in the past. However, the positive is we have been able to meet more often, a few times a year, to discuss items and keep moving issues forward.

BUSINESS CASE FOR CHANGE(S)

In late 2020 Peter put forward a Business Case for Change process for a restructuring of Forensic Chemistry. This is the new formal way of making changes within an organisation, which in the past would have occurred without much consultation.

The proposals were discussed many times and ultimately signed off, the outcome of which the instrument room and technicians became a separate standalone group, better able to serve the whole of forensic chemistry and was to be managed by a new

Senior Chemist position. Further, and based on practices occurring in forensic labs elsewhere, the illicit drug group has formed three separate teams for dealing with high throughput rapid cases, standard cases and complex cases to better meet turnaround time and provide more ability to deal with unusual drug items that come through.

In 2021, Queensland Health put forward a Business Case for Change across the whole department. Within it, our then current division, Health Support Queensland was scrapped and FSS was moved to the Prevention Division. Some 15-20 years ago, after the Ministerial Taskforce on Forensic Science reported on issues in providing forensic services to the State, FSS was separated from Pathology Queensland. This new move to the Prevention Division, once again reunited the two teams merged as a large entity. This has been the situation for the last 12 months; however, a recent update of the Queensland Health Business Case for Change has seen the Prevention Division scrapped and a new division created in which Pathology Queensland and Forensic and Scientific Services are separate once more. What goes around comes around!

Whatever division or format we are in, Forensic Chemistry has continued to provide excellent high standard forensic analysis and opinion-based evidence and the high calibre, professionalism, dedication, and real passion of staff to deliver the best they can, ensures Forensic Chemistry remains a great team to be a part of.

RIVER MONITORING

Long-time monitoring of the Burdekin River catchment (soil and water samples) suggests that there is low environmental, economic, and social risk of heavy metal remobilization from sugarcane in the Burdekin estuary during the dry season. Low concentrations and minimal enrichment factors associated with sugarcane suggest effective sugarcane practise (fertilizer application).

NON-ROUTINE SAMPLES

Routine samples are processed in a factory fashion. Nonroutine samples require more effort as the learning curve, which has made processing routine samples efficient, has not yet occurred. Non routine samples can identify problems that lead to government action. This leads to a resolution of the problems and a monitoring program to ensure that the problem does not recur. The non-routine becomes routine. Non-routine samples require a generalist chemist who is adaptable to different situations. Routine work requires a systems person who will follow the method religiously. Mixing them around can cause problems. A generalist may form the opinion that a step in the method is not necessary whilst a system person may not look too deeply into the method. In one case involving the determination of iron content of roofing paint, the method originally involved reduction of the iron then titrations. When instrumental techniques such as AAS came along the method was changed, the final determination being done by AAS. However, the point at which the method was changed was after the reduction of the iron 3 to iron 2. This was not necessary as the instrument is not concerned with the oxidation state of the iron as long as it is in solution. The generalist had a tough time convincing the systems chemist to change the method.

DR TATIANA KOMAROVA

Chief Chemist

Tatiana was born in a small Russian town Melenki Vladimir area. After finishing school, she moved to Moscow to study chemistry at Lomonosov Moscow State University. She completed master's degree and following PhD research in analytical chemistry dealing with Ion chromatographic determination of inorganic anions. That time Ion Chromatography was one of the novel and fast developing methods and Tatiana took part in the pioneering research in the field. Results were presented at the first International conferences on Ion Chromatography and competition of young scientists of Moscow State University in 1988 where Tatiana won one of the major awards.

Upon completion PhD Tatiana worked as a research scientist at the Vernadsky Institute of Geochemistry and Analytical Chemistry of Russian Academy of Science where in 1994 she was awarded with a National Scientific Scholarship for young scientists. She was very interested in research and development of ion chromatographic methods and with the help of her supervisor Prof. Yu. Zolotov established cooperation with a number of overseas laboratories and scientists including Prof. Paul Haddad from the University of Tasmania who was a well-known expert in the field. Tatiana decided to move to Australia to join Paul Haddad's group.

However, after arriving to Australia in 1995 she settled down in Brisbane. She met Prof. Barry Chiswell – the head of the Environmental chemistry laboratory at the Chemistry department of the University of Queensland who raised her interest in the research performed in his laboratory. In 1995 Tatiana was successful in her application for a Postdoctoral Research fellowship and joined Barry's group. She was involved in academic process, student supervision, research consultancy and development of new water treatment procedures using different analytical methods – HPLC, IC, FIA, AAS. The results of her post-doctoral research were implemented by several water treatment plants in the Gold Coast and presented at the International Congress on Analytical Chemistry in 1997 where Tatiana won the First Prize awarded by the Jury.

In 2002 after maternity leave she joined Queensland Alliance for Environmental Health Sciences (former National Research Centre for Environmental Toxicology) of the University of Queensland where she was involved in the development of innovative passive sampling techniques, managed chemical laboratory and participated in research consultancy work - environmental long-term monitoring of pollutants in different rivers feeding into the Great Barrier Reef's aquatic system with the use of passive samplers.

She continued research in passive sampling techniques when she moved to the Inorganic Chemistry laboratory of the Queensland Health Forensic and Scientific Services (QHFSS) in 2009. Diffusive gradients in thin films technique were successfully used in different national and international research consultancy projects and in several master's and PhD projects supervised by Tatiana. This work was recognised by the Queensland government as a finalist of the Premier's Award for Excellence in Public Service Delivery in 2011.

Working in the Inorganic Chemistry laboratory of FSS Tatiana has managed the Water Quality group performing analyses of drinking, dialysis and environmental waters for physical and chemical parameters which represent 75% of total laboratory business. She was also deeply involved in the development of new ICP-MS methods using Agilent Triple Quadrupole ICP-QQQ which were then NATA accredited and expanded laboratory capability and services. While working at QHFSS she completed a Diploma in Business, Management and Human Resource management at TAFE. This knowledge along with a PhD in analytical chemistry helped her to effectively manage the Inorganic chemistry laboratory while being appointed as Chief Chemist and Supervising Scientist at different stages of her career.

Tatiana's experience and expertise in chemistry have been well recognised - she is a Marquis Who's Who in the World listee since 2008 and a lifetime Marquis Who's Who in Science and Engineering listee, a member of International division of the Analytical Chemistry Council of Russian Academy of Sciences, a member of International Society of Environmental and Rural Development and Mercury Australia, a NATA technical assessor, Honorary Fellow of Queensland Alliance for Environmental Health Sciences (the University of Queensland) and adjunct Senior Lecturer in the School of Medical Science of the Griffith University. For many years she was a member of the QHFSS Research and Development Scientific Review Panel which on a competitive basis selected research projects for funding.

Tatiana is very passionate about research; its outcome was published in numerous peer-reviewed international journals and presented at different national and international conferences. Some of her publications and presentations were recognised by the International Society of Environmental and Rural Development, which gave her the award of excellent paper and award of sustainability promotion.

When Tatiana moved to the Inorganic Chemistry laboratory the majority of staff was busy with routine analyses and not interested in the research and development activity. With her sound research background and experience Tatiana did not feel happy in the laboratory until she received some support from Gary Golding, who was the managing scientist at that time. Gary helped Tatiana to get some funding for initial research projects and to promote her research outcome. Tatiana's enthusiasm paid off when she became a Chief Chemist over 60% of the laboratory staff were interested and involved in research and development activity. Gary Golding was well familiar with and very interested in the research work of Prof Yu. Zolotov, Tatiana's PhD supervisor, who is one of the world leading scientists in analytical and inorganic chemistry. Tatiana and Gary took an active part in organising Prof Yu. Zolotov's visit to Australia in 2006. He visited Brisbane, Perth, Melbourne, Adelaide, and Hobart, participated in a few conferences and RACI meetings, made eleven presentations, visited many laboratories and had numerous meetings and discussions with Australian scientists. Prof Zolotov describe his trip to Australia and analytical chemistry in Australia in a chapter "Analytical chemistry in other countries" of his book "Science about chemical analysis" published in 2023.

CHAPTER 25: EARLY NEWSPAPER REPORTS ON THE LABORATORY

THE GOVERNMENT ANALYST 1934

The Morning Bulletin Rockhampton 22 Sep 1934

In the past the press had a more positive attitude towards the laboratory.

"Stowed away on an upper floor of the Executive Building George Street, Brisbane, is the retreat of the government analyst, a place of grotesque glass globes, and sinuous tubes, of hair weight scales, and very matter of fact crucibles, flanked by meticulous measures and a prosaic whole family of the pestle and mortar series. Here, applied science unmasks applied acts, in its many guises, from water masquerading as milk, thermometers which breakaway from thermal standards and would, could they but agree thereon, set up standards of their own and enamel which gleams through an illegal film of Lead and Antimony. Here are to be found samples of oils and sardines, galvanised iron and some beverages, spirituous liquors and patent medicines, textiles, and chemicals. But if they vary widely – even grotesquely – in their nature, they are united in their mission. For this is the High

Court where the domestic history and molecular composition and standards of purity of all are alike brought under the x-ray, and from the decision of which there is no appeal. Last year the government analyst examined 12,217 samples, of which 6314 or more than half, were for the Health Department, so that a little known but wakeful eye keeps close watch on our physical well-being – so far as this is affected by pure foods – without many people being cognisant of the fact. Of the 6314 samples taken, 1152 or 18% – failed to pass the test, though in the case of disinfectants, toilet preparations, drugs, and medicines the reason for the failure was that the contents of the package did not justify the claims made on the label. Among the total samples for the Health Department were 2447 legal samples which were taken by inspectors in accordance with the provisions of the Health Act. Of these 127 were not in accordance with the provisions of the Health Act. Of these, 127 were fruit, 73% of which failed to get past the test, mainly on account of the presence of arsenic spray residues. "It is unfortunate for both public and growers that this dangerous practise of using poisonous insecticides cannot be stopped," says the Government analyst. "It affects the public because it renders the food poisonous. 60% of the samples of minced and sausage meat were below standard principally because of the use of sulphur dioxide as a preservative. Spirituous liquors and summer beverages which had fallen under strong suspicion of inspectors proved to be almost all below standard. The government analyst remarked, tritely, "adulteration with water was the only cause of failure in each sample which failed." By far the greatest proportion of samples examined for the Health Department, however, were milk and it is worthy of mention that only 15% of them failed. The analyst remarked that while the position still leaves plenty of room for improvement, particularly in some places outside the metropolitan area, it is much improved when compared with the average of the five years ending June 30, 1912. For the proportions of samples which then failed was 34%, and the proportion of added water was 12%. The average percentage of added water last year was 7.6%, with Sandgate 21.8%, Charters Towers with 12%, and Rockhampton with only 5% of added water."

PASSIVE SMOKING

Most blood and urine samples tested in the early 1980s tested positive for nicotine and cotinine (a nicotine metabolite), including children and sudden infant death cases. Many householders, commuters and workers were exposed to passive smoke before the laws were changed.

NOT SO PASSIVE - SMOKERS IN THE LAB

In the 1970s smoking in the laboratory was common. Extraction of aqueous liquids with ether was also common. The extracted aqueous phase with traces of ether was run directly into the sinks. One morning the deputy director tapped his pipe out into the sink. The ether exploded with flames shooting out of the sinks.

DEATH OF A CHILD

A magisterial inquiry was held at the Police Court on Wednesday afternoon, before Mr. E. T. Parr-Smith, J.P., to inquire into the death of a child named Hugh Roomer. The following evidence was taken:—Arthur Roomer said that he resided in Victoria-street, Bulimba; on the 4th January he resided in James-street, Fortitude Valley; on that morning his child Hugh died; he had been ill for some time, but he had not called in a doctor; witness gave the child some castor oil; he reported the matter to the police. Margaret Roomer said they had no means of obtaining a doctor's services; they came out in the Duke of Buccleuch on the 6th December. After Constable Murphy had been examined, Robert Marr deposed that there was no poison in the stomach and contents which he analysed. Dr. Wray said he made the *post-mortem* examination at the Morgue on the 4th January; the cause of death was, in his opinion, acute encephalitis; the organs were healthy; there were no external marks of violence. The inquiry, which was conducted by Senior-constable Fay, then closed.

COCAINE POSSESSION, SPRING HILL CAIRNS POST, 18 AUGUST 1933

POSSESSION OF POISON

COCAINE

A BRISBANE CASE.

BRISBANE, August 15.

A fine of £20, in default two months, was to-day imposed upon Cathleen Collins, 28, who pleaded guilty to having had a quantity of cocaine in her possession.

THE PLAINT.

The prosecution stated that the woman was intercepted at Spring Hill, and when her hand bag was searched two packets of cocaine were found. Defendant put one in her mouth and chewed it up. The other was sent to the Government analyst. The woman was known to be a cocaine addict and had also obtained supplies for other unfortunate women. Defendant denied that she had as-

Brisbane Courier, 22 October 1902 Possession/selling of a poison, an insecticide -

Government Analyst in Witness Box

Mr. Thomas McCall, Queensland Government analyst, spent yesterday afternoon in the witness box in the Police Court, being cross-examined most of the time about drugs when applied to tests.

He gave evidence in the case in which Alexander John Costello, of 127 Wickham Street, Valley, pleaded not guilty to having, under the registered firm name of the Parisian Medical Agency, sold a preparation containing ergot, otherwise than upon the prescription of a registered medical practitioner.

The hearing was adjourned until 11 a.m. on May 24.

CHAPTER 26: ACTIVITIES 2012 - 2022

The cutbacks of 2012, which involved a 27 percent (42 FTEs) reduction in chemistry staff, resulted in a consolidation of activities and the services offered by the chemistry laboratory. This involved a cessation of some services. Some traditional clients were very disappointed when they were told for the first time that the laboratory could no longer offer a particular service. We were not there when we were needed.

The Investigative Chemistry section, which specialised in low volume problem solving work, was closed with the loss of a chief chemist position and the remaining staff reallocated to the Forensic, Organic, and Inorganic sections. The Food Chemistry section was also closed with most remaining staff moving to Organic Chemistry, while others were transferred to Inorganic Chemistry. The Chief Chemist of Food Chemistry position was also abolished. The Foods Chemistry work was generally maintained though with fewer staff.

The remaining four chief chemists no longer reported to a chemistry discipline manager. The loss of a strategic chemistry managers with a thorough knowledge of the work of the chemistry laboratory will have a detrimental effect on the future. The routine factory laboratory can have a content free manager but a laboratory that solves client problems needs a manager with an understanding of the scientific discipline.

Without the monthly chemistry discipline meetings, the knowledge of the work of the other chemistry areas will fade with time. Transfers between sections cease. The generalist chemist will not be developed to the detriment of problem solving. This leads to a situation where the client decides the test required, much as the doctors do for pathology work, and the lab becomes a factory. One of the reasons for being is lost. Factory labs are the model used in the private sector. Problem solving gives the laboratory a competitive advantage needed to maintain budget funding. Decision making is slowed as the content free manager needs to be convinced often in writing for the most basic decisions. For example, in 2001 the turbo pump on a GCMS ceased to function. A content free manager could not immediately see the need to repair a work horse instrument, especially when it affected their budget.

COAL BOARD – REQUEST CHANGE OF RESULTS

An export coal sample failed the specification requirements and when this was reported to the Coal Board the person responsible asked that the result be changed so that the shipment passed. The chemist refused. It is interesting that the person from the Coal Board was in the merchant marine during the Second World War and once said he had been torpedoed and sunk twice in the Atlantic and

survived. I suppose this gave him a different perspective on what's important in life.

CHIEF CHEMISTS 2012-2022

Chief Chemist Forensic Chemistry: Peter Culshaw took over the management of Forensic Chemistry following the retirement of Chief Chemist Methsiri Edirisinghe.

Stewart Carswell and Mark Stephenson Chief chemists of Organics and Forensic Toxicology, respectively.

Eva Comino, Stephen Finlayson Chief Chemist

Inorganics

FORENSIC CHEMISTRY - WORK CHANGES
Provided by Dr Peter Culshaw

ILLICIT DRUGS

The number and diversity of novel psychoactive drugs continues to grow. Illicit drugs have continued to see new emerging drugs and have characterised many new ones of which staff have published their results in the scientific literature.

Of real concern was the emergence of derivatives of fentanyl, some of which are so toxic they are in the same hazard realm as chemical warfare agents. This led to new strategies being put in place to protect staff, including advanced first aid procedures incorporating training in the use of naloxone injection and nasal spray. Fortunately, Australia did not follow the same path as the US and Canada which has seen a large number of deaths of users of such drugs, and so the number of submissions of these substances and fatalities in this country has been extremely low to date.

Staff recognition of expertise within the illicit drugs group is highlighted by Senior Chemist, Karen Blakey being offered a part time role with the United Nations Office of Drugs and Crime.

CLANDESTINE DRUG LABS (CLAN LABS)

The clan lab team continues to have a close working relationship with the Queensland Police Service Illicit Laboratory Investigation Team (ILIT) by providing assistance to them upon request at suspected clandestine drug lab sites anywhere across the State.

Clan labs has continued to see new and emerging methods for the production of illicit drugs and staff have continued to carry out research to both confirm what is currently occurring in the field, and to proactively anticipate future potential emerging drug pathways.

The group took on their first PhD student in conjunction with Griffith University with

Dr Tim Currie a member of the team, taking on role of PhD supervisor. Recently, the team took on their first post-Doctoral research fellow from the University of Queensland to work on a collaborative project with law enforcement funding.

Staff within the group have also excelled personally in the past few years, including completing Master's Degrees in Forensic Science and a staff member now working on their own PhD research project, of which Peter is a co-supervisor.

EQUIPMENT

Whilst obtaining new staff has proven to be difficult over the last few years, we have been very fortunate to be able to gain a lot of new equipment and have been able to source many millions of dollars to replace a lot of ageing equipment. Sourcing new capital funding we have also acquired some new technologies, such as a benchtop Raman spectrometer, a microspectrophotometer, and a dual FTIR-Raman spectrometer, as well as a GC-solid phase FTIR.

TRACE EVIDENCE

Trace Evidence has provided the Queensland Police Service with a great deal of support training within the area of explosives investigation and are leaders in that area. Tony Peter, the team leader, has instigated some positive collaborations with both our Organics and Inorganics teams on site, especially in the use of isotope ratio mass spectrometry and ion chromatography research.

Trace evidence engaged an honours student to work on some research examining the comparison between various types of twine that had very positive outcomes.

LABORATORY INFORMATION MANAGEMENT SYSTEM (LIMS)

From mid-2016, Forensic Chemistry transitioned away from the Queensland Health based AUSLAB LIMS system to the QPS Forensic Register. This was a purpose-built forensic system designed by the QPS for their own use and was expanded to include access to Forensic Chemistry. It has a lot of great potential, and ultimately, we are looking to use such a system to move away from a paper-based system to a fully electronic case file.

INORGANIC CHEMISTRY - WORK CHANGES

Provided by Tatiana Komarova

August 2018 - Trace metal testing of clinical samples for Pathology Queensland (ISO 15189 accreditation) ceased. FSS needed to comply with the National Pathology Accreditation Advisory Council (NPAAC) requirements as part of its accreditation under NATA. The NPAAC required full-time on-site supervision by a Chemical Pathologist for clinical sample testing. This change degrades the role of the chemist in interpretation of results. It was not cost- effective for FSS to engage a Chemical Pathologist for the relatively small number of tests involved. However, the laboratory keeps ISO 17025 accreditation for trace metal testing of clinical samples for research and survey purposes. It is worth noting that in the past it was not uncommon for a doctor to ring the laboratory to ask a chemist for an interpretation of the level prior to going to court to testify. Chemists need to know this detail to target their analyses.

NEW WORK GAINED

Due to consolidation of activities the former Food Chemistry laboratory was separated into two sections based on the types of components analysed in food samples –organic and inorganic (trace metal) chemicals. The section performing trace metal analyses in food became a part of Inorganic Chemistry laboratory. Since then, analysis of trace metals in food samples has become a part of the services provided by the Inorganic Chemistry laboratory. Other new work included:

- NATA Accredited (ISO 17025, 2019) Passive sampling technique. DGT (Diffusive Gradients in Thin films) is used for sampling low levels of trace metals in different waters.

- Analysis of Lead in Paint – analysis brought by the Investigative.
- Still existing NATA accredited method for the Analysis of Lead and Cadmium in Crockery - analysis brought by the Investigative Chemistry staff was archived in 2018 due to insufficient sample numbers.
- Trace and Rare-Earth Elements in human blood by ICP-MS samples.
- Inorganics has established long-term cooperation with overseas commercial partners for the provision of metal speciation analyses.
- Capital expenditure on the laboratory has been made to make Inorganic Chemistry laboratory a biosecurity accredited facility, ideally suited for international collaboration.
- Inorganic Chemistry has membership with and is recommended by the Australian Mercury Society as a key contact for mercury speciation analysis supplements using an in-vitro digestion system.
- Trace metals in human scalp hair.
- Radiogenic Isotope and trace metal analysis of the Tully River
- Evaluation of mercury transfer from agriculture and mining in soil to freshwater and source identification.
- Analysis of environmental contaminants in Australia/Queensland honey.
- Lead testing in children, Townsville.

- Bioavailability study of trace elements and minerals from food formulars
- Fluoride in PFAS by ICMS
- Development of methodology for Strontium Isotope Ratio (87Sr/86Sr) analysis in water by ICP- QQQ without prior purification
- Application of ICP-MS for the analysis of nanoparticles in environmental and food samples
- Evaluation of rare earth and heavy metals transfer along the Burdekin River
- Development of new methodology for Total Phosphorus (TP) and Total Nitrogen (TN) by Segmented Flow Analysis (SFA)

RESEARCH DIRECTIONS AND OPPORTUNITIES:

About 60 percent of Inorganics staff is involved in research and development activities. Recent projects include:

- Assessment of maternal-infant status of essential trace elements in lactating mothers and their pre-term infants
- Influence of anthropogenic pollution on the level of toxic metals in blood
- Continue to develop analytical capability for the Dept Environment and Science for

- herbicide runoff and investigations of environmental harm.
- Formalised agreement with the Renal units for analysis of dialysis waters.
- Development of licit and illicit drugs in wastewater
- Species of metal compounds in food and biological samples
- Rare earth elements in water, soil, food, and biological samples
- Direct analysis of trace elements in sea water
- Lead and Strontium isotope ratios analysis in water and soil
- Nanoparticles in different samples by ICP-MS.
- Degradation products from waste treatment processes (e.g., fluoride in treated PFAS waste by IC-MS)
- Anions in specimens collected as legal evidence (acids, traces of explosives)

ORGANIC CHEMISTRY:

Provided by Stewart Carswell

The Organic laboratory continued to develop new technologies, for example,.

- Inorganics developed an increased capability to analyse for species of metal compounds in food.
- Organics acquired a new high resolution accurate mass liquid chromatography mass

spectrometer (Orbitrap LCMS) for the targeted and non-targeted analysis of residues of small molecule in environmental samples.
- Developed PFAS analysis in environmental and biological matrices.
- Other changes are towards automation or simplification of sample preparation throughput with same staffing levels
- Food Chemistry - allergen work and species identification by DNA is a newer area. This is a list of projects that generally develop capability.
- Organic Acids in Soils (with UQ) – Steve Carter.
- Characterization of potential toxic Blue Green algae from NE Australia (with DSITI) – Barbara Sendall
- Tutin and Mellitoxin in Honey – Kevin Melksham
- Beer and Cider IRMS study (Country of Origin) – Jim Carter
- Vitamins and Antioxidants – Hans Yates
- Development of methods for detection of potential cyanotoxins BMAA and DABA R Morgan
- Development of GCMS methods to detect and quantify regularly detected volatile organic compounds (VOCs) Jeff Herse
- Authentication of Australian honey Sadia Chowdhury, Shalona Anuj
- Licit and illicit drugs in wastewater Renu Patel, S Anuj, Steve Carter

- Characterisation of bloom-forming marine cyanobacteria in southeast Queensland by whole genome sequencing and toxin profiling Barbara Sendall, Amy Jennison, Rikki Graham
- Simultaneous analysis of selected food allergens in certain food
- The synthesis and characterisation of 3,4-methylenedioxyamphetamine (MDA) and 3,4- methylenedioxymethamphetamine (MDMA) from helional. Justin Cormick (Student), Jim Carter, Timothy Currie (supervisors)
- Isotopic characterisation of HMTD improvised explosives and hexamine precursors S Anuj, Tony Peter
- Developing a revised and enhanced methodology for the comparison of ropes, cords, and twine Tony Peter, Jim Carter, Shalona Anuj, and Jasper Bowman (Honours student) Authentication of Australian Stingless Bee Honey – proof of concept study Sadia Chowdhury
- Characterisation of potential anatoxin-producing cyanobacteria in southeast QLD and northern NSW Barbara Sendall, Glenn McGregor (DES), Paul Wright (Tweed Council)
- Development of position specific isotope analysis (PSIA) of amino acids as a potential diagnostic tool for metabolic change linked to cancer Jim Carter

- Evaluation of the Easy Plus Pro titrator for Food Chemistry spoilage methods Pam Kahlon, Sadia Chowdhury, Michael Geyer, Jim Carter

INVESTIGATIVE CHEMISTRY

Once the two streams of chemistry were combined to form the Inorganic and Organic chemistry sections in 1995, there seemed to be little scope for any further specialised sections to be developed. In the 1990's a section called Consumer Products became Manufactured Products and carried out testing on government contract items including fabrics used in hospitals, cleaning chemicals, floor polishes and institutional laundry chemicals. It also carried out control monitoring of paint samples for use on government buildings to ensure that the specified paint was used and that it had not been diluted. Over time the clients appeared to lose interest in this type of tender monitoring to ensure the quality of products provided to the government. They accepted the certificates from the manufacturer. Most of this work was subsequently suspended and the section moved in another direction. Sandy F very appropriately renamed the section Investigative Chemistry with a role in carrying out low volume analysis, drugs and poisons analysis, problem solving, chemical emergency analysis, introducing new techniques and providing service in chemical monitoring in the workplace, home, chemical emergencies, and the natural environment. Some of the achievements of this section included.

- The creation of a volatile organic compounds in air laboratory involving major purchases of equipment, the development of techniques, marketing, and provision of these techniques to government, universities, and industry. Many papers were published by the staff of this section in collaboration with various universities. This included a major survey of air quality in four major Australian cities.

- Our first international collaborative study yielded a perfect z score of zero. This was principally due to the use of internal standards on the thermal desorption tubes.

- Several PhD, Masters and honours students were supervised by staff.

- With the transfer of five health physics staff from Radiation Health, capability increased to include a service in contaminated site monitoring, radiation dosimetry and environmental monitoring. The staff were initially managed within Investigative Chemistry.

- The section was a testing ground for various Government initiatives including the commencement of Queensland Health's Performance Review process. The lab was able to make many suggestions to improve this process.

The closure of the Investigative Chemistry section following the redundancies of 2012 was a considerable loss to the Government's capability to deal with non-routine issues.

FORENSIC TOXICOLOGY

The work of this area differs from that of pathology laboratories in that the results must be beyond a shadow of doubt. This requires additional confirmation of the amount and identity of the drug. Presumptive tests are not sufficient. Regulations often have specific values above which the penalty changes e.g., 50mg per 100ml for blood alcohol. The analyst must be sure that the result is 51mg/100ml not 49 mg/100ml. Pathology results are interpreted against a range of concentrations.

The technology to carry out the testing has also advanced in leaps and bounds, especially during the past 50 years. Up until the 1980s thin-layer chromatography was the mainstay of screening techniques followed by confirmation by gas chromatography retention times. The introduction of electron capture and nitrogen phosphorus detectors on gas chromatographs further expanded capability. In the mid-1970s the first gas chromatograph mass spectrometer was purchased to enable the laboratory to detect tetrahydrocannabinol in drivers. In the following decades various mass spec techniques became readily available in benchtop instruments.

During the 1970s through to 2013 the toxicology area of the laboratory provided services to the coroner, the police, and corrective services.

In the 70s coronial samples consisted of stomach contents, livers, blood, and urine.

Most of the qualitative screening of samples was performed by thin-layer chromatography following an often-extensive solvent extraction process to separate the drugs into their groups e.g., acid drugs, base drugs. Urine samples were screened for drugs using EMIT, an enzyme technique, and thin-layer chromatography.

Don Lecky and later Geoff Rynja managed Toxicology during the 1970-1980. Warren Hamilton was supervisor of the laboratory for a period up to the mid-1990s. In the late 90s, Gary Golding was temporarily placed in charge of the area for two years. The environment was intensely political with most of the senior managers not having worked in chemistry let alone toxicology. Gary was not quite a content-free manager as he had previously worked in Toxicology. It was never made entirely clear to him why he was given this role in addition to his substantive position as supervisor of Investigative Chemistry.

The laboratory had long backlogs of samples which were directly related to the limited number of staff. The number of staff in the section in 2000 was about 14 people. Following efforts by senior management this was increased to over 32 people. This increase enabled turnaround times to be met. At the end of 2001 Neville Bailey was successful in applying for the position of chief chemist. Neville obtained his degree from the Queensland University of Technology in 1970. He was another one of Ivo Henderson's bonded scholarship holders, once again confirming the benefits of recruiting talented highly intelligent students prior to graduation by utilising a scholarship system. Neville spent most of his career in toxicology and had an unsurpassed depth of knowledge of the subject.

FUME-CUPBOARD FLOW RATE

One day the fume cupboards were going a lot faster than normal. Suddenly an evaporating dish was sucked up into the fume cupboard ducting. To improve the fume cupboard air flow in the William Street Toxicology lab, the State Works Department reversed pulley sizes on the fan and the resultant increase in flow blew rubbish that had accumulated in the pipes, over cars in the adjacent car park.

Toxicology obtained NATA accreditation for forensic work in 1998 and later in 2001, accreditation for the supply of reference materials, in this case alcohol solutions to calibrate the breathalysers.

Following Neville Bailey's retirement, Mark Stephenson took over the role of Chief chemist. Mark had worked in toxicology for most of his career.

During Gary Golding's period as chief chemist, toxicology acquired its first liquid chromatography tandem mass spectrometer. This technology opened the possibility of carrying out the analysis for a large range of drugs in blood in one instrument run, rather than the previous practice of identifying and quantifying drugs individually. The technique allowed for some automation of the analysis and data processing, however, it only found drugs that you were looking for and would not detect drugs that were not in the analytical suite. The Laboratory had for several years been using gas chromatography mass spectrometry to screen for drugs. However, there were issues with the thermal stability of some drugs. Liquid chromatography avoided this issue.

The forensic toxicology section was also responsible for carrying out the analysis of alcohol in blood samples taken from drivers by the police. In the 1960s alcohol was determined in blood by steam distillation and titration. This was very time-consuming. The GC technique involved pre-processing the blood sample to remove most of the protein matter. However, the process was simplified by dilution of the blood sample. The analysis was carried out by gas chromatography using a Flame Ionisation Detector (FID). The first GC auto injector ran off a program which consisted of holes in a plastic strip which actuated switches, similar to a pianola. Unfortunately, this strip often broke during overnight runs and the work had to be repeated. The lids on the autosampler glass vials were very difficult to put on and the vials often broke cutting the analyst's fingers.

The laboratory worked closely with the police to develop technology to carry out roadside saliva testing for drugs in drivers. For the first time roadside cannabis tests were possible. The police carried out roadside testing and if positive, obtained a sample of saliva which was sent to the laboratory.

Samples of urine were sent to the laboratory by various agencies which had an employee "no drug use" policy. Other urine samples were obtained from the Corrective Services and the Drug Courts. Over time as usually occurs, new management at Corrective Services decided they would like to outsource the work to get a cheaper deal rather than rely on the more expensive government laboratory. This eventually led to the loss of this work. Following the cutbacks in staff in 2012, there was an organisational restructuring which placed the toxicology lab under the management of forensic pathology. This imitated the Victorian Institute of Forensic Medicine (VIFM). The VIFM was itself attached to the coronial courts which were co-located.

This made decisions on the amount of testing relatively easy as the coronial decision-makers and the analysts were in the same building.

SHIPPING INCIDENTS

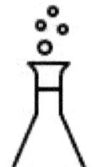

Periodically, incidents occur on the ocean near to the coast. On one occasion containers of ammonium nitrate were washed off a ship into Moreton Bay. Tests looked at contamination of the bay with these environmental nutrients. On other occasions, ships spilled oil and contaminated the beaches north of Brisbane and in Gladstone. Commercial fishing had to stop until the possibility of contamination had been eliminated by the laboratory.

THE LOTTERY

Before a compute generated random numbers for the lottery, discs were used. These discs were placed into a machine and rotated. At a certain point a number of discs would drop into a slot on the machine., The job of the laboratory was to ensure that there was no bias in the discs by examination and weight. This work disappeared when a computer generated the numbers.

CHEMISTRY ANNUAL CHRISTMAS PARTY

For most of its history on the final workday before Christmas the chemistry laboratory held celebration lunch. They invited all the old retirees. At his funeral, Noel Long's (instrument maker), family specifically mentioned how he appreciated the annual invitation to the work Christmas party, even after 25 years. This practice generated a certain esprit de corps. Unfortunately, there were always staff who objected to subsidising the retired staff attendance. Others preferred to go home early to take care of their own Christmas preparations. Some of the early parties, when the staff consisted predominately of young males, in their early 20s, were very noisy. One year, prawns were on the menu. There were balloons filled with hydrogen resting on the high ceilings of the food laboratory. Somebody decided that a PVC tube with a wad of wet tissue and a prawn head could be used as a blow pipe, to burst these balloons. Not many balloons survived.

On one occasion the section, whose turn it was to organise the party, ordered a large quantity of spring rolls from a local Chinese restaurant. When they were delivered, they were frozen. There were always hitches! One administrative officer Theresa Capper, who had a talent for catering, always did a good job organising these events when her section was delegated the job.

As the gender balance changed many of the women had to make their own preparations for their family Christmas and were not interested in a work party. The Forensic DNA were not interested in joining the custom. Since the laboratory no longer invites retirees or does not have a laboratory party, the retired staff still get together each year at a local tavern. The retired staff invite the existing staff.

TOUCH POWDER

In early 1970s a practical joke was often perpetrated with nitrogen triiodide. Nitrogen triiodide is made by mixing small quantities of aqueous ammonia solution with an ethanolic iodine solution. A precipitate forms, which when dry, is an extremely sensitive explosive. When placed on the ground while still wet and allowed to dry, it will explode with a loud pop when trodden on. Unfortunately, it leaves a brown iodine stain which can be removed with sodium thiosulphate. Never make large quantities as it is extremely unstable.

ENVIRONMENTAL HEALTH OFFICERS AND THE LAB

Historically there has been a close association between the environmental health officers and the laboratory. In the public health areas, the EHO'S brought in the samples for testing as part of their regulatory role. As successive governments have cut back on checking and moved to a deregulation approach to safety, fewer and fewer samples were consequently brought in.

Many of the chemists and the environmental health officers got to know each other when they delivered samples to the lab. Many had relationships dating back decades. With the introduction of central sample receival this personal contact was lost. EHO David Logan was one of the long-term supporters of the lab. He managed the Brisbane South EHOs. He mentioned to Gary Golding one day that his lease on the building they were in near Garden City was running out. Gary Golding saw an opportunity to house them at the laboratory. He advised David to ring the CEO and see if that was possible. The CEO supported the move. They became the only EHOs the lab staff knew personally. Centralised sample receival reduced lab staff contact with other EHOs.

The change in the role away from regularly checking compliance with regulations to periodic checking via a special project meant that the steady stream of samples previously received ceased. Committees consisting of lab staff and EHOs were set up to run these projects.

At the start of the covid pandemic an EHO was on duty at the border when Gary Golding re-entered the country. They recognised each other and the EHO said "I used to bring samples to you at the lab".... disturbingly past tense.

STAFF WAGES IN CASH

Fortnightly, in the 1970s, a government car was sent down to the Reserve Bank in Adelaide Street to collect the payroll for the staff in cash. This cash was sorted into envelopes by the head clerk and the staff lined up to sign for their pay. Once electronic payments came, the whole process was much more efficient and reliable, having fewer errors than the manual system and was not subject to armed holdups. The staff still had to line up, this time at the bank, to withdraw cash for weekly expenditures!

PLATINUM CRUCIBLES AND DISHES

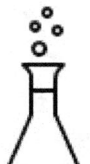
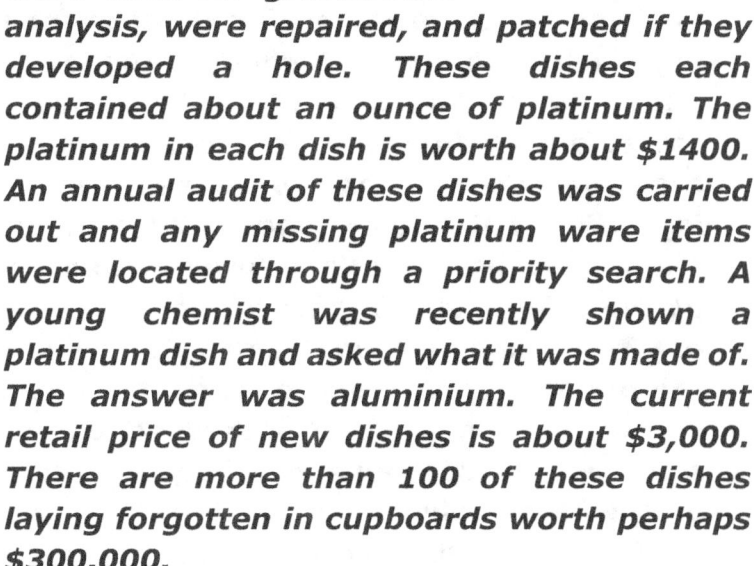

Platinum dishes and crucibles, which were used for gravimetric analysis, were repaired, and patched if they developed a hole. These dishes each contained about an ounce of platinum. The platinum in each dish is worth about $1400. An annual audit of these dishes was carried out and any missing platinum ware items were located through a priority search. A young chemist was recently shown a platinum dish and asked what it was made of. The answer was aluminium. The current retail price of new dishes is about $3,000. There are more than 100 of these dishes laying forgotten in cupboards worth perhaps $300,000.

SICKLY COCKTAIL

A lady tourist drank a Gin Sling cocktail at a hotel, but after drinking only a small amount she began vomiting violently and felt very ill. Her husband thought she had been poisoned, and called the police, who submitted the remains of the drink and all the

bottles of ingredients that had been used to make it. All those bottles of ingredients had been used previously without adverse effects, although Gin Slings were infrequently prepared. Analysis showed the drink contained very high levels of zinc, which is a powerful emetic when ingested. The hotel measured the ingredients for the drink using jiggers attached to the bottles. The type of jigger used refilled with the ingredient at the same time as a measure of the ingredient was dispensed, so the ingredient sat in the jigger until next used. The jiggers were made of brass with chrome plating. The fruity ingredient contained a lot of citrus juice and was acidic. As this sat in the jigger for prolonged periods, the internal chrome plating gradually dissolved, exposing the brass. Brass is an alloy of copper and zinc and will dissolve in acid. but the acidity preferentially dissolved the zinc, causing the problem. Many years ago, a similar problem occurred with oysters from the Derwent River in Tasmania near a zinc refinery.

 SUFFOCATION AT SEA

A young sea cadet was found dead in the toilet of a ship. He had suffered prolonged sea sickness and had been vomiting in the toilet, where he was *found, apparently suffocated. Investigations revealed that the sewage tank on the ship was fitted with a goose-neck water trap through which the gases from the tank (including methane and carbon dioxide) would vent as pressure built up. The trap was filled with sea water. Examination of the trap revealed that it was completely blocked with a white solid, causing the gases from the tank to flow back into the toilet bowl, where they concentrated, since they were heavier than air. The cadet had inhaled sufficient of the gases to become unconscious and finally suffocated.*

Analysis of the white solid showed that it was mostly magnesium carbonate with some calcium carbonate. Over time the carbon dioxide in the sewage gas had reacted with the calcium and magnesium in the sea water to produce the insoluble carbonates, which eventually blocked the vent. Although sea water contains much more calcium than magnesium, magnesium carbonate is much less soluble than calcium carbonate and was deposited preferentially.

CHAPTER 27: RECOGNITION OF CHEMISTRY STAFF AND FORMER STAFF

Tatiana Komarova - Premier's award - Women in Science

Lieutenant Colonel R. A. Stanley DSO, VD

John Brownlie Henderson - Royal Australian Chemical Institute – Fellow, Order of the British Empire

David Grantham - Order of Australia for work in occupational safety

Stewart Bell - Royal Commissioner - New Zealand mine explosion

Gary Golding - Australia day medallion from Qld Health, Royal Australian Chemical Institute - Distinguished Fellow - Australia day medallion from Dept of Emergency Services, RACI Qld branch president, Order of Australia Medal for contribution to Science - Chemistry

Ron Biltoft - Australia day medallion from Qld Health

Graham King - Australia day medallion from Qld Health, RACI Qld branch president

Stewart Carswell - Australia day medallion from Dept of Emergency Services - Royal Australian Chemical Institute – Fellow, RACI Qld branch president

Henry Olszowy Australia day medallion from Qld Health

Scott Turner – Fire and Emergency Services medal

Peter Culshaw – Royal Society of Chemistry – Fellow, A/Professor Griffith University

Mary Hodge - Royal Australian Chemical Institute – Fellow, Australia day medallion from Qld Health

Michelle Neil - Royal Australian Chemical Institute - Fellow

Geoff Rynja - Royal Australian Chemical Institute - Fellow

Pieter Scheelings - Royal Australian Chemical Institute - Distinguished Fellow

Graeme White A/Professor Bond and Griffith Universities

Paul Pisasale – Mayor of Ipswich

These are only the people I am aware of at the time of publication. The list demonstrates the calibre of the lab people.

MICROBIOLOGICALLY STERILE PRAWNS

An incident of note involved the results on some cooked imported prawns which tested as 'microbiologically sterile'. The prawns originating from Thailand via the Netherlands were suspected as having been irradiated which would make them a prohibited import. Subsequent analysis at Melbourne University's Electron Spin Resonance facility, which can detect residual free radicals in solid matrices such as bone and exo-skeletal issue, confirmed that the prawns had indeed undergone irradiating treatment likely in the Netherlands which has several food irradiation facilities. No further consignments of sterile cooked prawns were subsequently received.

CHAPTER 28: IMPROVING THE CHEMISTRY LABORATORIES

In writing this book several concepts have come to mind as a result of 150 years of experimentation with different systems of management. The following suggestions are made with a view to increasing the capability and survivability of the chemistry laboratory and to avoid repeating past mistakes.

LEADERSHIP - SENIOR MANAGEMENT

Appoint a permanent, non-contract executive director who can provide long-term leadership. The commission of inquiry into the DNA lab may not have happened if the lab had had a permanent long-term leader.

Since Greg Shaw left in 2015, there has been a stream of people acting or short-term appointments to the position of executive director. This eight years of lack of an experienced, permanent committed leader, has been detrimental to the future of the laboratory as demonstrated by the DNA inquiry which would not have occurred if a permanent leader had dealt with the issues. Leadership of a laboratory is more complex than leadership of a government regulatory or policy area. It involves a learning curve, scientific background, management skills, strong problem-solving skills, an ability to work in a bureaucratic system, media skills and long-term commitment. These skills may only be found in people who have worked within the laboratory. Internal appointments should be considered.

STRUCTURE

- combine all the chemistry groups under one manager who has chemistry qualifications – recruit from within.
- change structure to include middle managers – this will allow time for strategic initiatives and time to listen to staff – the progressive flattening of the structure from five senior managers to one has been detrimental - the flat structure has not been successful - this expansion of structure will provide promotional opportunities which do not exist in a flat structure – succession training will occur automatically – promotion from within the organisations will again become possible.

RECRUITMENT

- run a scholarship system through the Rural Health Scholarship Scheme to support high achieving students studying chemistry to ensure the laboratory recruits the best people – bond them - leave positions vacant during the year to ensure there is a position

to place these people into - offer two scholarships per year!
- annually run a recruitment process to provide a short list that can be drawn upon at any time there is a vacancy without having to wait every time for a long-drawn-out recruitment process to commence.
- be sure on their first day that new staff are introduced to senior management whose job is to ensure the values of the organisation are communicated

STAFF DEVELOPMENT

- ensure all staff have a project to work on which will result in the development of additional capability including admin and support services - potentially these projects could become part of a PhD
- encourage staff to carry out additional formal studies.
- encourage staff to join the RACI to improve networking, their project management and
- presentation skills - support their efforts – most past senior managers were members
- make an MBA or at a minimum a Graduate Diploma in Management an essential requirement for those seeking promotion to management positions – why would you hire managers who have not studied the subject any more than you would hire a chemist who has not studied chemistry

- develop complex report writing skills.
- get people out of the laboratory into the real world to assist in solving problems
- develop generalist chemists who can put a hand to a wide range of problems – this will involve transferring staff between sections - solving major problems is a key competitive advantage within government
- all public health staff should be familiar with forensic procedures e.g., continuity of possession and security of the sample
- introduce a continuing professional development scheme as a pre-requisite for salary increments like that which exists in pharmacy
- undertake and act upon regular staff well-being surveys to establish staff mental and physical health
- undertake and act upon staff exit surveys to identify and address barriers to staff development and career advancements

THE ADVANTAGES OF ONE GOVERNMENT LABORATORY

There are considerable overheads in running a major laboratory. These include administration, management, procurement, training systems, implementing and maintaining a quality system, sample charging systems, and ensuring sufficient instrument capability. The cost of the latter is reduced by eliminating the duplication of expensive instruments. New technology activities are centralised so that specialised units service the entire organisation. Management of IT and quality must be duplicated in each organisation.

Spreading the laboratory across different departments complicates the pay scales for staff. Can you remain on the health practitioner scale after a move to a different department as proposed for the Forensic sections e.g., to the Justice Department? In 2012 the Health Practitioner's salary scale went two levels higher than the Professional Officers scale (PO).

Invariably, conflict arises between staff members. It is useful to separate the combatants by moving one to a different section. This tool is not easily used if the staff member must be moved between different Government Departments.

The question must be asked "How does the work of the laboratory support the mission of the department they work in. One of the reasons given for the expansion of law schools in universities is that the law course has low overheads compared with science. They do not use expensive infrastructure or expensive equipment. Movement of the forensic areas to the Department of Justice will face this problem of dealing with a department that is not used to budgeting for expensive equipment. The skills of lawyers in debating issues will put the scientists at a disadvantage when arguing a case.

LABORATORY INFORMATION MANAGEMENT SYSTEMS

- run two systems, one that can be easily adjusted for changes in work, client, analytes etc without the cost of involving the vendor and a second system that covers routine work and does not need to be adjusted

MARKETING

- publish stories in the press – in the past 20 years I have only heard complaints about the labs - take the initiative.
- Periodically check that the lab maintains its status as an Australian Defence Department strategic resource.
- Get a shorter name so the press can use it.

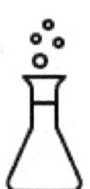

VOLATILES FROM FLOORING

As society changed and confidence in government turned to suspicion, the closeness of the laboratory to government became an impediment on some occasions. For example, a school in Northern Queensland was experiencing problems with volatile organic compounds following the installation of a new resin-based flooring system. At the last-minute, the laboratory was required to withdraw from the investigation because of perceived bias. Unknown to the participants in this debate, this avoided an even bigger problem as the technician assigned to visit the school and take the samples, turned out to be the son-in-law of the builder of the school. The work was given to a private consultant from Melbourne.

TRACE ELEMENTS IN HONEY

Inorganics provided the first data on mineral levels in Queensland/Australian honey. The study has shown wide variations of elemental levels in Australian honey products, and in particular, the high levels of K and Zn. The study has shown that honey could contribute a significant dietary source for these elements. Relatively low levels of trace elements including toxic heavy metals were found in the study, and the values are comparable to other countries. The study also found significant differences between urban and rural honeys for some elements (B, Na, P, Mn, Sr), but no trend.

CHAPTER 29: SOME USEFUL MANAGEMENT TOOLS

BUILD TEAMS

Chemists work in teams, lead teams, and cooperate with other teams. As such, understanding team dynamics, the development of teams and the issues facing a team is critical. Here are a few thoughts on this subject.

BE AWARE OF THE STAGES OF TEAM BUILDING.

It has been established through team research that teams move through a series of stages.

FORMING: this happens when the team first comes together, and everybody is sensitive to the views of others and are trying to find their place in the team.

STORMING: To establish acceptable codes of conduct and the methodology of the team operations there is a stage of vigorous argument about these factors. This is a very stressful time for team members but is a necessary part of the process of establishing behaviour, that is, what is acceptable and what is not acceptable. Being aware of this stage can help team members accept any criticism without becoming too emotional. If the leader attempts to bypass this stage by threats conflict may arise during the performing stage to the detriment of performance.

NORMING:
The team establishes standards of behaviour in the way the team does its work.

PERFORMING: in this stage the team carries out the work for which it was formed.

TERMINATING: The work of the team has been completed and people start to cut their ties with the team. There may be celebrations for the completion of the work.

Care must be taken to allow the stages to progress in that order. Often there is a temptation among the team Leader to regard the storming process as bad conduct and to suppress the opinions of subordinates, however research has shown that if a team leader forces members to by-pass the storming process, difficulties arise in the performing stage when it is desirable for the team to function smoothly.

You may find that during the initial meetings of the team, when the storming process has commenced there are people in the team who you feel you cannot possibly get along with only to find at a later stage after the norming process is completed so that they are quite cooperative and helpful. I suspect that at the storming stage people in roles of authority may be seen as bullies for arguing against the views of some of the other team members. It is helpful to reduce the stress in the team building process caused by the storming process for what it is, just a stage. Don't get upset by it, just work through it. If the storming process continues there are major problems for the team and other actions are needed. Looking at the team roles currently present in the team. For example, your team may get bogged down in an issue because there is nobody on the team who is a good problem solver. This leads us to Belbin Team roles.

BELBIN TEAM ROLES (more detail www.Belbin.com)

Meredith Belbin's research showed that the most successful teams were made up of a diverse mix of behaviors. To build high-performing teams, we need to represent each of the nine Belbin Team Role behaviours at the appropriate times. The nine Belbin Team Roles are: Resource Investigator, Team worker and Co-ordinator (the social roles); Plant, Monitor Evaluator and Specialist (the Thinking roles), and Shaper, Implementer and Completer Finisher (the Action or Task roles).

It may help to think about the original Star Trek crew.

Captain James T. Kirk: Shaper

Spock: Monitor Evaluator

Dr. Leonard "Bones" McCoy: Team worker

Montgomery "Scotty" Scott: Specialist

Nyota Uhura: Resource Investigator

Hikaru Sulu: Implementer

Pavel Chekov: Completer Finisher-

The following is taken from **BELBIN.COM**

RESOURCE INVESTIGATOR:

Uses their inquisitive nature to find ideas to bring back to the team.

Strengths: Outgoing, enthusiastic. Explores opportunities and develops contacts.

Allowable weaknesses: Might be over-optimistic and can lose interest once the initial enthusiasm has passed.

Don't be surprised to find that: They might forget to follow up on a lead.

TEAM WORKER:

Helps the team to gel, using their versatility to identify the work required and complete it on behalf of the team.

Strengths: Co-operative, perceptive and diplomatic. Listens and averts friction.

Allowable weaknesses: Can be indecisive in crunch situations and tend to avoid confrontation.

Don't be surprised to find that: They might be hesitant to make unpopular decisions.

COORDINATOR:

Coordinator Needed to focus on the team's objectives, draw out team members and delegate work appropriately.,

Strengths: Mature, confident, identifies talent. Clarifies goals.

Allowable weaknesses: Can be seen as manipulative and might offload their own share of the work.

Don't be surprised to find that: They might over-delegate, leaving themselves little work to do.

PLANT:

Tends to be highly creative and good at solving problems in unconventional ways.

Strengths: Creative, imaginative, free-thinking, generates ideas and solves difficult problems.

Allowable weaknesses: Might ignore incidentals and may be too preoccupied to communicate effectively.

MONITOR EVALUATOR:

Provides a logical eye, making impartial judgements where required and weighs up the team's options in a dispassionate way.

Strengths: Sober, strategic, and discerning. Sees all options and judges accurately.

Allowable weaknesses: Sometimes they lacks the drive and ability to inspire others and can be overly critical.

Don't be surprised to find that: They could be slow to come to decisions.

SPECIALIST:

Brings in-depth knowledge of a key area to the team.

Strengths: Single-minded, self-starting and dedicated. They provide specialist knowledge and skills.

Allowable weaknesses: Tends to contribute on a narrow front and can dwell on the technicalities.

Don't be surprised to find that: They overload you with information. Specialists are common in scientific organisations. This may lead to a deficit of other roles

SHAPER:

Provides the necessary drive to ensure that the team keeps moving and does not lose focus or momentum.

Strengths: Challenging, dynamic, thrives on pressure. Has the drive and courage to overcome obstacles.

Allowable weaknesses: Can be prone to provocation and may sometimes offend people's feelings.

Don't be surprised to find that: They could risk becoming aggressive and bad-humoured in their attempts to get things done.

IMPLEMENTER:
Needed to plan a workable strategy and carry it out as efficiently as possible.

Strengths: Practical, reliable, efficient. Turns ideas into actions and organises work that needs to be done.

Allowable weaknesses: Can be a bit inflexible and slow to respond to new possibilities.

Don't be surprised to find that: They might be slow to relinquish their plans in favour of positive changes.

COMPLETER FINISHER:
Most effectively used at the end of tasks to polish and scrutinise the work for errors, subjecting it to the highest standards of quality control.

Strengths: Painstaking, conscientious, anxious. Searches out errors. Polishes and perfects.

Allowable weaknesses: Can be inclined to worry unduly, and reluctant to delegate.

Don't be surprised to find that: They could be accused of taking their perfectionism to extremes on-routine issues."

TRACE ELEMENTS IN BLOOD AND PLASMA

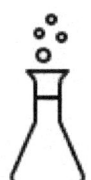

Trace element analysis in whole blood and plasma for reference levels in a selected Queensland population. There has been no recent study investigating a wide range of trace elements in the Queensland population from various regions and among age groups, sex, and health status. Our work is the first study providing the reference range values of essential and non-essential trace elements in blood and plasma of a selected cohort of adults in Queensland. The study has filled a current knowledge gap and provided important information for health workers. The data could be used as a guideline for health assessment of toxic metal exposure, evaluation of nutritional status and diagnosis and screening of certain diseases.

COMMISSION OF INQUIRY INTO DNA LABORATORY 2022

"Analytical results, quick, cheap, or right choose any one"

Gary Golding 1994.

Stories appeared in the press around a murder case where no DNA was found in blood samples and there were conflicting results between similar samples. The accusation was made that criminals were being set free after the trials because of lack of DNA evidence. There was a view that the detection limit had been set so high that profiles were being missed. The Premier ordered a commission of inquiry into the DNA laboratory. In simple terms, the laboratory was diluting samples to a higher degree than in other states. This reduced the probability of a profile match. The Commissioner also found that the reports issued were not truthful. The laboratory reports stated there was insufficient DNA to produce a profile but in fact there may have been. It was left to the police to ask for further testing.

In the early days of the 1990's, DNA technology was being developed at the laboratory. The laboratory developed their own database. When the national database was developed, the DNA laboratory was forbidden from using their own local database.

As the value of the evidence became apparent more and more samples were sent to the laboratory, resulting in a huge backlog of tens of thousands of samples. There was almost daily press criticism and criticism by magistrates.

Up until the early 2000's the technicians would take the samples and carry out the whole process. This was not effective for many samples. A production line approach was implemented as one of the managers said "when you break the process down into smaller steps anybody can do the work. The flaw in this approach was that nobody had a full overview of the entire case so any information which might have shed light on the interpretation of results was not available.

The commission of inquiry investigated the DNA laboratory, its processes, validation of methods and the general work environment. The result of this multimillion-dollar investigation was that it was recommended that the forensic labs be transferred out of Health into the Justice Department as a more appropriate place for them to be housed. Whether this will come to pass remains to be seen as the Justice Department may have a different view.

The result of the transfer of forensic chemistry and DNA analysis to the Justice Department will mean a breakup of the laboratory, environmental and public health chemistry may remain in health even though they also do legal analysis of food and environmental samples. The development of generalist chemists through transfer between departments will be much more difficult. As a consequence of this lack of ability to move staff around, over time will be a lack of knowledge of the capabilities of the other areas. It will also result in duplication of equipment. Past experience of transfers of staff to other departments have already been discussed. Specifically, the transfer of the occupational health laboratory to the Department of Workplace Health and Safety resulted in a closure of the laboratory and transfer of the funds and the staff to management and inspectorial positions. This is unlikely to happen in the DNA case, as the police will be very vocal should there not be any results. The concept of having a single laboratory servicing all government analytical needs will cease. There will be an expansion of administrator positions due to the duplication of requirements in the different departments.

ARE THEY SCALLOPS OR STINGRAY

One noted example included DNA verification of sting-ray flesh fashioned and labelled as scallops on sale in Brisbane. Laboratory staff also featured in several Current Affairs programs relating to the meat content of meat pies and the mislabeling of restaurant fish meals.

As part of a gradual expansion of technical expertise, areas of interest included work on the development of ciguatera in coral trout led by Dr Ian Stewart, the chemical profiling of beverages particularly wine and beer of suspect origin and, with the acquisition of a state-of-the art IRMS and the recruitment of Dr J Carter which facilitated the provenance determination of imported foods and beverages.

SOMETIMES THE JOB STINKS

In the 1960's the large freezer containing toxicology specimens of livers, stomachs and other biological material broke down over an Easter break. By Tuesday morning when staff returned the stench was extreme. As junior members of staff, Ron Biltoft and Doug McGregor were given the task of cleaning out the freezer, (which was quite deep) so it could be repaired. The smell permeated their hair and clothing, ensuring other people gave them plenty of space. Because the samples were all from legal coronial cases they could not be discarded, but still had to be analysed. The putrefaction products in the tissues made the analyses of drugs much more difficult, as well as unsavory. The fume cupboards at that time were very inefficient, and bad odours wafted through the building for many weeks until all the putrid samples were completed. Complaints were even received from DPI staff two floors above the Toxicology laboratory.

CHAPTER 30: OTHER GOVERNMENT LABORATORIES

PIETER SCHEELINGS, PHD FRACI

"Experience is the name everyone gives to their mistakes."

Oscar Wilde 1854 – 1900

EXPERIENCE AT THE AUSTRALIAN GOVERNMENT ANALYTICAL LABORATORY (AGAL)

Pieter started his professional career at the Commonwealth Customs Laboratories in Melbourne in January 1973. While public records are somewhat scant, the Customs laboratories were created well before federation and were located in all major ports to collect excise on imported rum often valued as an alternative currency in addition to imposing tariff duties on imported products to protect local manufacture. After federation, the Commonwealth Customs Department assumed jurisdiction for each of the various state Customs Laboratories other than Queensland where the state laboratory continued to undertake customs work on behalf of the commonwealth. Over the years, the role of Customs laboratories was expanded to control of dangerous substances, the determination of authenticity of products and the provision of scientific advice relating to tariff and related duties on imported goods. Gradual reductions in international trade barriers diminished the professional role of the tariff chemist which, due to the complex nature of tariff classification, had generally been recognised as a career pinnacle in government chemistry laboratories.

Pieter Scheelings commenced his employment in the Drug Research Section of the renamed Australian Government Analytical Laboratories in Melbourne which was still under the jurisdiction of the federal Customs department. His initial contract was in a postdoctoral position before appointment to research chemist and finally to Section leader. As part of an international collaborative study on cannabis, the primary role involved studies on the pyrolysis of cannabis and cannabis extracts and the synthesis of cannabinoid homologues to assist in the chemical profiling of cannabis imports. Other functions included a study on determining the tar and nicotine content of cigarette smoke using a prototype cigarette smoking machine on loan from Monash University. This preliminary work was in advance of the federal government legislating for the labelling of tar and nicotine levels on cigarette packaging. The study was extended to estimating the tar content of 'in-house manufactured' cannabis cigarettes and an international collaborative project on the tar content of Thai tobacco as part of a cancer study on indigenous Thai smokers. The duties also included the commissioning and application of the first GC/MS (Varian MAT 111) employed in Australian government laboratories, particularly for identifying illicit drug seizures by Customs. The expertise of the Drug Research Section was acknowledged when staff were invited to co-author a research paper titled Some Recent Advances in the Study of Cannabis (G.B Chesher, R. Malor and P. Scheelings) for the 1979 S.A. Sackville Royal Commission into the Non-Medical Use of Drugs.

In 1986, Pieter transferred to Adelaide to take up the role of Regional Director of the Adelaide AGAL laboratory where he developed an increasing interest in food analysis. The Adelaide laboratory had developed a reputation for expertise in food composition and nutritional analysis due to its contractual work in a number of national FSANZ food analysis programs including the annual market basket surveys and the nutritional composition of commercial and ready-to eat foods for inclusion into the Australian Food Composition Tables. These programs involved the purchase and preparation and cooking of foods to reflect consumer food intake prior to analyses. Other major programs included analysis of imported foods, the National Heart Foundation 'Pick-the-Tick' program and the export certification of Australian wines to the EU. The laboratory's microbiology group were the first national laboratory to develop and validate a range of microbiological assays for water-soluble vitamins to complement the chemistry methods for vitamins in foods. The small microbiology group received some public notoriety when it first reported the presence of listeria in several soft cheeses in a small survey of imported cheeses after similar results had been reported in the EU. While the results were 'of some interest' to the Government Analyst, the lab was advised that such surveys were the primary responsibility of the Health and Customs departments which appeared to be put 'offside' by the laboratory's staff initiative.

With the closure of the Adelaide laboratory in 1997, Pieter resigned from AGAL to take up the position of Team leader, Food Chemistry at FSS, Coopers Plains.

LOSS OF ANALYTICAL EXPERTISE WITHIN GOVERNMENT LABORATORIES

The restructuring and downsizing of state and commonwealth analytical laboratories in the mid-nineties due in part to the introduction of 'fees-for-services' and in part to various governments' responses to waste and mal-administration initially reported by the Ross Inquiry into Commonwealth Laboratories resulted in significant loss of technical, professional and support staff to the analytical chemistry community.

The closure of the SA State laboratories in 1995, the closure of two AGAL regional laboratories in 1997 and the downsizing of the Victorian State Chemistry Laboratory resulted in an estimated loss of some 200 chemistry-based staff coupled with an associated loss of corporate knowledge and scientific expertise from government agencies. The unintended consequence of the commercial focus imposed on government laboratories often resulted in significant program redirection from traditional government programs to high volume and more profitable environmental work. The use of sample compositing with some loss of analytical sensitivity and the introduction of shift work to deal with higher workload and shorter client turn-around times was associated with an increased incidence of quality breakdowns. The more rigid public service employment conditions and bureaucratic management structures in time provided private laboratories with a considerable competitive advantage.

CAREER OBSERVATIONS AT GCL/QHFSS

The first week at Coopers Plains was a pleasant revelation. In the first week the laboratory held an R&D workshop where several GCL and ENTOX staff presented summaries of their current and planned research work which indicated a strong R&D culture. Another early observation was that there appeared to be a more relaxed approach to commercial activities. This mirrored the early days at AGAL and appeared more consistent with the traditional role of government laboratories, i.e., analytical services and associated advice to government agencies. It became clear that the research/problem solving culture, and the promotion of postgraduate students and training of international scientists was, in part, due to the 'academic approach' that the current and previous director brought the GCL. supported by GCL senior management, APFAN as a host and vehicle for training food scientists from developing countries had been in operation for some 8 years under the coordination of Dr Howard Bradbury and effectively managed by Graham Craven. By 1997, 4 training workshops had been completed at Coopers Plains with some 80 or so trainees graduating from these workshops. In addition, a few staff had conducted training workshops in Indonesia led by Dan Wruck and Geoff Rynja. These 'training' activities could not have been undertaken as commercial ventures. With ongoing support from QHFSS and funding from several external agencies including the Crawford Fund, AusAID and ACIAR, APFAN managed a further 3 practical workshops at Coopers Plains as well as organising several international conferences. During this time Howard Bradbury passed on the role of

APFAN Coordinator to Pieter Scheelings. (See cartoon)

The work of the Food Chemistry Section involved classical food composition analysis for protein, fat, sugars, dietary fibre, metals, and preservatives for compliance with national and state food standards to ensure consumer safety and truth of labelling primarily on behalf of the Department's Environmental Health Unit. A more specialist function included the analysis of food complaint samples which covered a broad range of physical and chemical contaminants requiring considerable experience from the relevant State Analysts Neil Douglas and Inge Scott. In response to an increasing incidence of allergenic reactions by consumers to specific foods, the Section expanded its service capabilities to include analysis for a range of allergens using ELISA-based methodologies. The secondment of molecular biologist Annette Baddeley to the Food Section facilitated the application of DNA technology to determine the presence of GMO ingredients in foods and replacing the less reliable iso-electric focussing protein analysis by DNA sequencing for the confirmation of fish and other species. The appointment of Glen Graham enabled the setting up of a dedicated molecular biology unit within Food Chemistry to facilitate the routine application of DNA methods for identification of undeclared or mis-labelled fish and meat species in imported and retail foods.

During his employment at FSS, Dr Scheelings maintained an involvement in professional societies including his role as chair of the RACI Employment and Emolument Committee, representation of the RACI at annual NATA Council meetings, co-chair of the RACI 1998 congress in Adelaide and membership of state RACI analytical chemistry Groups. He was appointed to the editorial board of the journal Managing the Modern Laboratory, was a member of the AOAC Technical Committee on Reference Materials and a member of the Australian delegation to the annual meetings of the Codex Alimentarius Committee on Pesticide Residues where he represented the delegation on the Methods of Analysis sub-Committee taking on as Chair from 2011-12 . In recognition of his services to Queensland Health and professional societies, Pieter was awarded a QH Australian Day Medallion, the 2012 AOAC Reference Materials Achievement Award, several RACI citations and, in 2022, was appointed a distinguished fellow of the RACI for services to the Institute including his 50-year membership. He has been an external examiner of several post-graduate theses, chaired three SAA committees and represented QH on several interdepartmental committees providing technical advice of food chemistry methods. He has undertaken several externally funded reviews of international laboratories on behalf of Queensland Health including Mozambique (AusAID), Papua New Guinea (ACIAR), Vietnam (UNIDO) and, early in his career, government laboratories in Saudi Arabia. Pieter retired in 2013 after 40 years of service in government Laboratories but has continued to provide some technical support for NATA food laboratory accreditation assessments.

CHAPTER 31: THE FUTURE
"The
future is a convenient place for dreams."
Anatole France 1884 to 1924

Having served the community and Government for 150 years, the laboratory has proven its worth to the State. This service has been supported by successive Governments. This should be considered by those who seek to implement the latest management fad. Its involvement across a variety of Government departments and constant references to its analyses in the press indicates a need for a continuation of the service. The increase in the variety of scientific services offered is of immense value to the State in solving problems that constantly arise in our complex society. The labs' role in foreseeing these problems, generating data to enable decision making and bringing this to the attention of the relevant responsible authorities increases its value. In the previous 70 years prior to the 1990s, appointments to the position of director were invariably from within or from a closely related government laboratory.

This demonstrates the in-house development of future managers. People were selected by a tap on the shoulder to move to an area several years before the incumbent departed to learn the job. The hierarchical structure enabled the opportunity for several prospective managers to be developed at the one time, increasing the chance that one of them would demonstrate the talent to take over the retiring director's position. The words "succession training" were never heard, it just happened. Once the structure was flattened with the subsequent removal of middle management positions there were fewer opportunities to understudy the single incumbent. Outside appointments of people with little or no knowledge of the business became common. These people tended to focus their attention on budget management and not on the development of the laboratory. Greg Shaw went against this trend by appointing one deputy director and all four managing scientists from within the organisation. The deputy director left and was not replaced. The following period was generally regarded by the chemistry staff as a happier period free of formal commission of inquiries, task force investigations, and interdepartmental inquiries that plagued the laboratory in the period before and after Greg Shaw's tenure.

This complex would probably not exist if it had not been for the vision of Trevor Beckmann. The structure represents a significant investment and probably saved the laboratory from the fate of being sold off, as has happened to similar public health labs in South Australia and the UK.

The Covid 19 pandemic highlighted the shortcomings of outsourcing everything to the private sector and the value of having a multidisciplinary, centralised, government controlled, laboratory complex, which fulfils the mandate.

"BE THERE WHEN NEEDED

BUT MAKE SURE THEY KNOW YOU ARE THERE"

"The final test of a leader is that he leaves behind him in other men the conviction and the will to carry on."

Walter Lipman 1889 – 1974

 **FINALLY, WE RETURN TO TOILET PAPER**

A lady complained that a roll of toilet paper had produced a burning sensation when used by several members of her family. The paper developed slightly brown areas, which analysis showed to be due to the presence of sulphuric acid. It appeared that the paper had been contaminated with battery acid during transport.

APPENDIX 1: ACRONYMS

AAS Atomic Absorption Spectrometry
AFFIN Australian Future Forensics Innovation Network
AFP Australian Federal Police
AGAL Australian Government Analytical Laboratories
ALS Australian Laboratory Services
ANU Australian National University
ANZFSS Australia and New Zealand Forensic Science Society
ANZPAA Australia New Zealand Policing Advisory Agency
AOAC Association of Official Analytical Chemists
APFAN Australian Pacific Food Analysis Network
AusAID Australian Aid
CRC Centre for Collaborative Research

CSIRO Commonwealth Scientific and Industrial Research Organisation
DAF Department of Agriculture and Fisheries
DEEDI Department of Employment, Economic Development, and Innovation
DFAT Department of Foreign Affairs and Trade
DPI Department of Primary Industries
EHO Environmental Health Officers
ENTOX Environmental Toxicology (University of Queensland)
FAAS Flame atomic absorption spectroscopy
FACS Fluorescence activated cell sorting
FID Flame Ionisation Detector
FSSA Forensic Science South Australia
FTIR Fourier Transform Infrared Spectroscopy
GC Gas Chromatography
GCL Government Chemical Laboratory
GCMS Gas Chromatograph Mass Spectrometer
NHMRC National Health and Medical Research Council
GFAAS Graphite furnace atomic absorption spectroscopy
HPLC High Performance Liquid Chromatography
ICP Inductively coupled plasma
ICPAES Inductively coupled plasma atomic emission spectroscopy
ICPMS Inductively coupled plasma mass spectrometry
NIFS National Institute of Forensic Science (Australia)
NIRA National and International Research Alliances program
NMI National Measurement Institute
NRCET National Research Centre for Environmental Toxicology

PAFP Partnership-Alliances Facilitation Program
PC (laboratory) Physical containment
QH Queensland Health
QHFSS Queensland Health Forensic and Scientific Services
QHPSS Queensland Health Pathology and Scientific Services
QHSS Queensland Health Scientific Services
QIS Quality Information System
QIT Queensland Institute of Technology (later QUT)
QPS Queensland Police Service
QUT Queensland University of Technology
RACI Royal Australian Chemical Institute
TATP Triacetone triperoxide
TGA Therapeutic Goods Administration
UQ University of Queensland
UTAS University of Tasmania
UTS University of Technology Sydney
VIFM Victorian Institute of Forensic Medicine
XRF X-ray fluorescence

APPPENDIX 2: AUTHORS BACKGROUND

The author, Gary Golding, commenced working at the Queensland Government Chemical Laboratory on a permanent basis in December 1971 and continued to work at the laboratory until he retired in September 2013. During this time he worked in the public health, environmental, toxicology and forensic areas. At the end of his career, he was the Managing Scientist of all the chemistry sections of the laboratory employing over 140 professional chemists.

His formal qualifications included, a Master of Applied Science (Analytical Chemistry) and a Master of Business Administration (Technology Management). He is a Fellow of the Royal Australian Chemical Institute (RACI) and was awarded a Distinguished Fellowship in 2020. He served two years as the Queensland branch president of the Institute. In 2024 he was awarded an Order of Australia Medal for contributions to science – chemistry.

He wrote this book to encourage students to seek a career in chemistry, provide advice to chemists and laboratory managers, and for the general public who wonder what chemists do.

APPENDIX 3: PUBLICATIONS AND CONFERENCE PRESENTATIONS

The following list of publications and oral presentations by the author demonstrates the breadth of interesting work involved in a career in analytical chemistry even in a non-research environment.

1) **Ketamine Abuse** W. R. Pease and G. M. Golding Proceedings of the Australian Forensic Science Society vol. 2 1982 pp. 23-28.

2) **Blood Lead Levels in Queensland Children** M. Rathus, S. Leatham, G. Golding, C. Rowan Medical Journal of Australia vol. 2, 1982 pp. 183-185.

3) **Talc: an Unusual Component in Cannabis Resin** G. M. Golding and W. R. Pease Microgram: A publication of the United States Drug Enforcement Administration September 1986.

4) **The Applications of Fourier Transform Infrared Spectrometry at the Government Chemical Laboratory** G. M. Golding Proceedings of the Royal Australian Chemical Institutes seminar on FTIR, Brisbane October 1987.

5) **The Thin Layer Chromatography of Textile Dyes: Data Maximisation** G. M. Golding and S. Kokot Proceedings of the Tenth Australian International Forensic Science Symposium, Brisbane 1988.

6) **The Selection of Non-correlated Solvent Systems for the Thin Layer Chromatographic Separation of Dyes Extracted from Transferred Fibres** G. M. Golding and S. Kokot Journal of Forensic Sciences vol. 34, no. 5, 1989 pp. 1156-1165.

7) **A Simple Device for the Reduction of Background in FT-IR Instruments** G. M. Golding and B. Davis Applied Spectroscopy vol 43, no. 4, 1989 pp.

8) **Comparison of Dyes from Transferred Fibres by Scanning Densitometry** G. M. Golding and S. Kokot Journal of Forensic Sciences vol. 35, No. 6, 1990 pp. 1310-1322.

9) **Applications of Thin Layer Chromatography in Forensic Chemistry** G. W. Lee, L. Hadley and G. M.

Golding Proceedings of the 10th Australian Analytical Conference Brisbane 1989.

10) **The Rapid Screening and Determination of Toxic Metals in Post Mortem Liver Specimens** G. M. Golding and P. Geoghegan Government Chemical Laboratory Report Series No. 6 1992.

11) **Expiry Dates and Vitamin Preparations** G. Tempany, C. Malar, S. Nilsen, K. Reardon, G. Golding The Australian Journal of Pharmacy Vol. 74 September 1993

12) **Contract Quality Assurance Analysis** G. Craven, G. Golding, G. Tempany, G. King, C. Malar Proceedings of the conference of the Australian Cosmetic Chemists Association. Gold Coast September 1994

13) **Chemical Emergencies - Waste Management by the Road Side** G. Golding Proceedings of the Royal Australian Chemical Society Seminar on Waste Management Brisbane 1997

14) **The use of chemometrics and FTIR to investigate the chemical and mechanical degradation of cotton during the washing process** was presented in Darwin at the 13th

Analytical Chemistry Conference 1995. This poster paper was the result of a collaboration with Serge Kokot, Chemistry Department QUT

15) **Hypothetical - The future of chemical education** Proceedings of the International Conference on Chemical Education, University of Queensland July 1996

16) **The Analysis of Volatile Organic Compounds in Air** G. Golding Commonwealth Science Council Workshop on Air Quality Brisbane July 1999

17) **Volatile Organic Compounds: Relevance and Measurement in Australia Indoor air quality** literature review and research survey G. Golding and E. Christensen A publication of Environment Australia October 1999

18) **Chemical analysis at chemical emergencies** G. Lee and G. Golding Commonwealth Science Council Workshop on chemical emergencies Malaysia November 2000

19) **Chemical Emergency response – protecting the community and the environment** G Golding International Union of Pure and Applied Chemists IUPAC 38th World chemistry Congress Brisbane 2001

20) **Project Management - managing the human factor** G. Golding Project Management workshop Brisbane 2001

21) **Chemical emergency response** Invited speaker Australian Radiation Protection Society ARPS26 Gold Coast September 2001

22) **Chemical Emergency response – The role of the analytical chemist** G. Golding Interact 2002 chemistry conference Sydney July 2002

23) **Characterisation and Identification of Sources of Volatile organic Compounds in and Industrial Area of Brisbane** O. Hawas, D. Hawker, A. Chan, D. Cohen, E Christensen, G. Golding and P. Vowles Proceedings of the CASANZ Conference August 2002

24) **Sampling and Analysis of Ambient Volatile Organic Compounds (VOCs) in an industrial area in Brisbane Australia**. O. Hawas, D. Hawker, A. Chan, D. Cohen, E Christensen, G. Golding and P. Vowles, Journal of Clean Air and Environmental Quality Vol. 36 No.4 Nov. 2002 p40-45

25) **Fine Particle Composition in Four Major Australian Cities – Elemental Composition of Aerosols** Hawas, O., Stelcer E., Cohen D., Button, D., Denison, L., Wong N., Chan,A., Simpson, R., Christensen, E., Golding, G., Hodge, M., Kirkwood, J., Mitchell, R., Wainwright, D. & Ardern, F. (in press). 17th International Clean Air & Environment Conference, November 2003 Newcastle.

27) **Occupational Hygiene Outlook for Modern Health Care** D. Grantham, J. Wright, G. Golding, G. King, G. Lee, B. Wallace Australian J. of Occup. Health and Safety Feb 2003

28) **A Pilot Survey of the Presence of Undeclared drugs and Health Risks Associated with Metal Contamination of Complementary Medications offered for Sale in Queensland** S. Rutherford, I. Marshall, A. Loan, G. Tempany, G. Golding Environmental Health Vol 3 No. 4 2003 p. 21 28

29) **Apportionment of Sources of Emissions of Particles in Four Major Australian Cities by Positive Matrix Factorization** Andrew Chan, Rod Simpson, David Cohen, Olga Hawas, Eduard Stelser, Lyn Denison, Neil Wong, Gary Golding, Elizabeth Christensen, Willy Gore5, Mary Hodge, Eva Comino, Stewart Carswell 18th International Clean Air & Environment Conference, May 2005 Hobart.

30) **Trace Heavy Metals in Fine and Coarse Aerosols in Four Major Australian Cities** Olga Hawas1, Eduard Stelcer, David Cohen, Andrew Chan, Rod Simpson, Lyn Denison, Neil Wong, Gary Golding, Elizabeth Christensen, Willy Gore, Mary Hodge, Eva Comino, Stewart Carswell 18th International Clean Air & Environment Conference, May 2005 Hobart.

31) **Probabilistic Risk Assessment as a Framework to Assess Public Health Risks Associated with heavy metals and metalloids in Chinese Herbal Medicines.** Cooper, K.J. Noller, B., Connell, D., Yu J., Sadler, R., Rutherford, S., Olszowy, H., Golding, G., Tinggi, U., Analytical Chemistry Conference 1995. This poster paper was the result of a collaboration with Serge Kokot, Chemistry Department QUT

Analytical Chemistry Conference 1995. This poster paper was the result of a collaboration with Serge Kokot, Chemistry Department QUT

15) **Hypothetical - The future of chemical education** Proceedings of the International Conference on Chemical Education, University of Queensland July 1996

16) **The Analysis of Volatile Organic Compounds in Air** G. Golding Commonwealth Science Council Workshop on Air Quality Brisbane July 1999

17) **Volatile Organic Compounds: Relevance and Measurement in Australia Indoor air quality** literature review and research survey G. Golding and E. Christensen A publication of Environment Australia October 1999

18) **Chemical analysis at chemical emergencies** G. Lee and G. Golding Commonwealth Science Council Workshop on chemical emergencies Malaysia November 2000

19) **Chemical Emergency response – protecting the community and the environment** G Golding International Union of Pure and Applied Chemists IUPAC 38th World chemistry Congress Brisbane 2001

20) **Project Management - managing the human factor** G. Golding Project Management workshop Brisbane 2001

21) **Chemical emergency response** Invited speaker Australian Radiation Protection Society ARPS26 Gold Coast September 2001

22) **Chemical Emergency response – The role of the analytical chemist** G. Golding Interact 2002 chemistry conference Sydney July 2002

23) **Characterisation and Identification of Sources of Volatile organic Compounds in and Industrial Area of Brisbane** O. Hawas, D. Hawker, A. Chan, D. Cohen, E Christensen, G. Golding and P. Vowles Proceedings of the CASANZ Conference August 2002

24) **Sampling and Analysis of Ambient Volatile Organic Compounds (VOCs) in an industrial area in Brisbane Australia**. O. Hawas, D. Hawker, A. Chan, D. Cohen, E Christensen, G. Golding and P. Vowles, Journal of Clean Air and Environmental Quality Vol. 36 No.4 Nov. 2002 p40-45

25) **Fine Particle Composition in Four Major Australian Cities – Elemental Composition of**

Aerosols Hawas, O., Stelcer E., Cohen D., Button, D., Denison, L., Wong N., Chan,A., Simpson, R., Christensen, E., Golding, G., Hodge, M., Kirkwood, J., Mitchell, R., Wainwright, D. & Ardern, F. (in press). 17th International Clean Air & Environment Conference, November 2003 Newcastle.

27) **Occupational Hygiene Outlook for Modern Health Care** D. Grantham, J. Wright, G. Golding, G. King, G. Lee, B. Wallace Australian J. of Occup. Health and Safety Feb 2003

28) **A Pilot Survey of the Presence of Undeclared drugs and Health Risks Associated with Metal Contamination of Complementary Medications offered for Sale in Queensland** S. Rutherford, I. Marshall, A. Loan, G. Tempany, G. Golding Environmental Health Vol 3 No. 4 2003 p. 21 28

29) **Apportionment of Sources of Emissions of Particles in Four Major Australian Cities by Positive Matrix Factorization** Andrew Chan, Rod Simpson, David Cohen, Olga Hawas, Eduard Stelser, Lyn Denison, Neil Wong, Gary Golding, Elizabeth Christensen, Willy Gore5, Mary Hodge, Eva Comino, Stewart Carswell 18th International Clean Air & Environment Conference May 2005 Hobart.

30) **Trace Heavy Metals in Fine and Coarse Aerosols in Four Major Australian Cities** Olga Hawas1, Eduard Stelcer, David Cohen, Andrew Chan, Rod Simpson, Lyn Denison, Neil Wong, Gary Golding, Elizabeth Christensen, Willy Gore, Mary Hodge, Eva Comino, Stewart Carswell 18th International Clean Air & Environment Conference, May 2005 Hobart.

31) **Probabilistic Risk Assessment as a Framework to Assess Public Health Risks Associated with heavy metals and metalloids in Chinese Herbal Medicines.** Cooper, K.J. Noller, B., Connell, D., Yu J., Sadler, R., Rutherford, S., Olszowy, H., Golding, G., Tinggi, U.,., Moore, M.R., and Myers, S. First World Congress on Chinese Medicine Abstracts 21-24 November 2003 Melbourne Town Hall, Melbourne pp:168-169.

32) **Development of a Probabilistic Risk Assessment Approach to Assess Public Health Risks associated with heavy metals and metalloids of Traditional Chinese Medicines.** Cooper, K.J. Noller, B., Connell, D., Yu J., Sadler, R., Rutherford, S., Olszowy, H., Golding, G., Tinggi, U., Moore, M.R., and Myers, S. Oral presentation. Queensland Health Scientific Meeting 25-26 November 2003 Brisbane Convention Centre, Brisbane. pp1.

33) **Public Health Issues associated with heavy metals and metalloids in Chinese Herbal Medicines and the Role of Probabilistic Risk.** Cooper, K.J. Noller, B., Connell, D., Yu J., Sadler, R., Rutherford, S., Olszowy, H., Golding, G., Tinggi, U., Moore, M.R., and Myers, S. Australian Pharmaceutical Science Conference, Sydney 3-5 December 2003.

34) **Quantitative GC-MS analysis of Δ 9-tetrahydrocannabinol in fiber hemp varieties.** Hewavitharana AK, Golding G, Tempany G, King G, Holling N. Journal of Analytical Toxicology 2005;29(4):258

35) **Terrorism Response – Issues Involved in the Development of a Chemical Weapons Analysis**

Capacity within State Public Health and Forensic Laboratories Gary Golding, John Bates, Richard Mattner, Graham King, Tony Peter, Ross Kleinschmidt, Ross Sadler, Graeme White Interact 2006 RACI conference Perth WA 2006

36) **Terrorism Response – Issues Involved in the Development of a Chemical Weapons Analysis Capacity within State Public Health and Forensic Laboratories** Gary Golding, John Bates, Richard Mattner, Graham King, Tony Peter, Ross Kleinschmidt, Ross Sadler, Graeme White The 6th Annual Health and Medical Research Conference of Queensland 2006

37) **Public health risks of traditional Chinese medicines.** Cooper, K., Noller, B., Connell, D., Sadler, R., Ozslowy, H., Rutherford, S., Golding, G. and Tinggi, U. (2002) Posters Queensland Health and Medical Scientific Meeting, 4-5, December, 2002, Brisbane.

38) **Public Health Risks of Traditional Chinese Medicine** Cooper, K., Noller, B., Connel, D., Sadler, R., Olszowy, H., Rutherford, S., Tinggi, U. and Golding, G. (2003).. Posters General Practice and Primary Health Care Conference. 18th-20th June 2003. National Convention Centre, Canberra. 142-143.

39) **Pilot study on Lawn Mower VOC Emissions** Tapper S., Christensen E., Soon Chee Chan, Golding G., Sadler S Poster Interact 2006 RACI Conference Perth WA

40) **Potential for exposure of airport fire crews to volatile organics during training exercises** Tapper S., Christensen E., Soon Chee Chan, Golding G., Sadler S Poster The 6th Annual Health and Medical Research Conference of Queensland 2006

41) **Pilot study on Lawn Mower VOC Emissions** Tapper S., Christensen E., Soon Chee Chan, Golding G., Sadler S Poster The 6th Annual Health and Medical Research Conference of Queensland 2006

42) **The Role of Paracetamol (acetaminophen) in the Reduction of Tremor in Parkinson's Disease – a Case Study,** Golding G. M. (2020) *British Journal of Pharmacy.* 4(2). doi: https://doi.org/10.5920/bjpharm.619

www.ingramcontent.com/pod-product-compliance
Lightning Source LLC
Chambersburg PA
CBHW071015240426
43661CB00073B/2292